Displacing Territory

Displacing Territory

Syrian and Palestinian Refugees in Jordan

KAREN CULCASI

The University of Chicago Press
Chicago and London

The University of Chicago Press, Chicago 60637
The University of Chicago Press, Ltd., London

Published 2023
Printed and bound by CPI Group (UK) Ltd, Croydon, CR0 4YY

32 31 30 29 28 27 26 25 24 23   1 2 3 4 5

ISBN-13: 978-0-226-82704-9 (cloth)
ISBN-13: 978-0-226-82706-3 (paper)
ISBN-13: 978-0-226-82705-6 (e-book)
DOI: https://doi.org/10.7208/chicago/9780226827056.001.0001

Library of Congress Cataloging-in-Publication Data

Names: Culcasi, Karen, author.
Title: Displacing territory : Syrian and Palestinian refugees in Jordan / Karen Culcasi.
Description: Chicago : The University of Chicago Press, 2023. |
Includes bibliographical references and index.
Identifiers: LCCN 2022059266 | ISBN 9780226827049 (cloth) |
ISBN 9780226827063 (paperback) | ISBN 9780226827056 (ebook)
Subjects: LCSH: Refugees—Jordan. | Refugees—Syria. | Refugees—Palestine. |
Refugees—Legal status, laws, etc.—Jordan. | Refugees—Jordan—Attitudes. |
Human territoriality—Middle East. | Palestinian Arabs—Jordan—Social conditions. |
Syrians—Jordan—Social conditions. | Middle East—Boundaries—Public opinion. |
BISAC: SOCIAL SCIENCE / Refugees | SCIENCE / Earth Sciences / Geography
Classification: LCC HV640.4.J6 C85 2023 | DDC 362.8709569—dc23/eng/20230208
LC record available at https://lccn.loc.gov/2022059266

⊗ This paper meets the requirements of ANSI/NISO Z39.48-1992 (Permanence of Paper).

Contents

Figures

Acknowledgments

Ali al-Asmar and Dima Obeidat assisted me in my research in Jordan. They were both amazing assistants whose work was crucial for my research. I was fortunate to receive several grants for this research, including a fellowship from the West Virginia University Humanities Center, a National Science Foundation ADVANCE grant, and an Eberly College of Arts and Science ARTS grant. Brenden McNeil, my colleague, husband, and partner in crime, supported me intellectually and emotionally throughout the years of researching and writing this book. Ayla and Milo's frequent queries of "Hey, Mom, how's the book going?" proved to be the little extra encouragement I needed to get it done.

Susannah Engstrom, at the University of Chicago Press, along with two anonymous reviewers, provided me with invaluable recommendations on the flow and substance of my arguments. Tamara Ghattas, also at the University of Chicago Press, was a generous and patient copyeditor who had countless insightful comments and suggestions for improving my manuscript. I'd also like to thank Meagan Walker for help with making the maps in this book, as well as refreshing my rusty ArcGIS skills.

I have immense gratitude for the people I met and talked with in Jordan over the course of many years. My research participants taught me so much about the life and politics of refuge, as well as about humanity and generosity. I thank them so much for their hospitality and for teaching me about the complexity of their lives.

I will donate a portion of the proceeds from this book to one particular aid organization in Jordan working with refugees. In an attempt to help correct some of the wrongdoings of US foreign policy in Iraq, two American women formed the Collateral Repair Project (a name derived from the military term "collateral damage," which refers to civilian casualties) in 2006. This NGO works to help "repair" the damage caused by the American military invasion of Iraq that ultimately led to the mass displacement of Iraqis to Jordan. You can learn more about their important work at https://www.collateralrepairproject.org/.

Introduction

Refugee Crises

In 2014, there were 59.5 million forcibly displaced people across the globe. Of these people, 8.3 million were *newly* displaced that year, which was the highest ever yearly increase of displaced people since records have been kept (UNHCR 2015). Stemming mostly from conflicts and persecution in Syria, Afghanistan, and Iraq, the large uptick in forced displacement in 2014 affected those states and the neighboring states of Turkey, Lebanon, and Jordan most directly and intensely. Yet the Western world's attention focused on the effects of these forced displacements in and near Europe. As the number of displaced people continued to increase in 2015, many European states panicked as just over a million displaced people sought asylum via the Mediterranean Sea in Greece and Italy and another 34,000 displaced people walked over land routes from Turkey to Bulgaria and Greece (Clayton and Holland 2015). Soon dubbed the "European refugee crisis," the expansive, multifaceted, and grave situation was discursively construed as a crisis experienced by Europeans and European states, not by the asylum seekers themselves, nor by the states of Turkey, Lebanon, and Jordan, which were managing millions more new refugees. While 1,034,000 asylum seekers is certainly a large number, the European Union's population in 2015 was 508,200,000 (Eurostat 2015), which means that these displaced people constituted a minuscule 0.2 percent of the total population of the EU.

There were immense variations in how European states responded to these large numbers of asylum seekers. Some states—like Sweden and Germany—allowed refugees in, worked to ensure their safety, and made earnest attempts to treat them humanely. But other states—like Italy and Hungary—created and expanded actions to purposely deter and restrict the entry of displaced people into their state-territories (Bose 2020; De Genova, Garelli, and Tazzioli

2018; Jones 2016; Barnett 2003; Squire 2020). Leveraging the fear of "crisis" (Conlon and Hiemstra 2017; Culcasi, Skop, and Gorman 2019) and fueled by anti-immigration sentiments that constructed refugees as harbingers of trouble and danger (Holmes and Castañeda 2016; De Genova 2018), deterrence measures such as building fences near borders and policing the sea to stop boats worked to keep displaced people from crossing into the EU to seek asylum. In other words, several European states used the discourse of a "crisis" to justify securing their borders in order to stop new refugees from entering their state-territories. Further, these exclusionary practices occurred despite the fact that all EU states are party to international laws and agreements that guarantee the right of displaced people to seek asylum. The prioritization of securing state borders over helping asylum seekers created what De Genova, Garelli, and Tazzioli (2018, 240) refer to as "countless real crises" for those people seeking safety. In great part because of these deterrence measures, many displaced people—people who had already experienced the trauma of living through war, fleeing their homes, and undergoing arduous journeys to seek safety—were subjected to renewed death and despair.[1] For example, four thousand people drowned in the Mediterranean Sea in 2015 alone; seventy-one people died of apparent suffocation in an abandoned truck in Austria in August 2015; and countless other refugees have suffered in overcrowded refugee camps that lacked the most basic provisions, such as in Calais, France. Europe's "refugee crisis" brought to bear the tragedy that can evolve when states ignore international laws and refuse the entry of asylum seekers. Of course, such practices are not unique to a few states in Europe. Australian and US administrations have echoed a similar discourse about the threat of refugees, implemented violent border practices to deter them, and created their own state-centered crises in the process (Little and Vaughan-Williams 2017; Agnew 2019).

The conflicts occurring in Syria, Afghanistan, and Iraq that led to the 2014–15 mass displacements had more direct and more intense ramifications in the states of Turkey, Lebanon, and Jordan than in Europe. Each of these three states struggled to manage the massive influx of refugees and certainly experienced some degree of economic and humanitarian "crisis" as a result. Yet they each managed far larger numbers of refugees, with fewer resources and less tragedy and death, than within Europe.

Jordan is a resource-poor, Global South state that has had its independence only since 1946. Throughout its entire seventy-six years a state, Jordan—officially, the Hashemite Kingdom of Jordan—has accepted several million displaced people fleeing conflict in neighboring countries. As a result, as of early 2021, Jordan has the second-largest refugee population per capita in the world (after Lebanon). This includes 664,414 registered Syrian refugees and an estimated 600,000 unregistered Syrians who all fled to Jordan since the Syrian war erupted in 2011 (ACAPS 2022). In 2014 and 2015, when Europe was experiencing its "refugee crisis," Jordan was overwhelmed by the sheer numbers of Syrians seeking asylum within its territory. Yet the Syrians who sought safety in Jordan were not systematically turned away, nor were they widely subjected to anti-refugee rhetoric or vilified as threats. After a year of maintaining open borders and accepting Syrians in the hundreds of thousands, the Jordanian government altered these welcoming practices and policies and began to restrict Syrians from crossing the border as well as forcing Syrians into refugee camps. These newly implemented exclusionary practices did not, however, fully replace Jordan's more open and inclusive practices that had existed for many decades.[2] Instead, inclusionary and exclusionary measures have mixed together, creating an incredibly complex refugee landscape across Jordan.

In this book I delve into the varied policies, practices, and experiences of forced displacement in Jordan since its independence in 1946, through detailed case studies of Syrian and Palestinian refugees. While I focus on Jordan's specific practices and the experiences of Syrians and Palestinians, I situate my study within the broader context of *global* refugee policies and practices. Ultimately, by highlighting the complexity of Jordan's refugee system and by humanizing refugees who have experienced forced displacement, my goal is to critique the discourse of "refugee crises" that are commonly decried in the Global North.

There are many factors and concepts that matter deeply in examining refugee policies and practices—like national identity, economics, and geopolitics. And while I recognize all these factors, I focus pointedly on the concept and context of territory. Territory is a foundational concept that frames countless international and state policies and practices related to forced displacement. Territory is also central to displaced people's experiences while fleeing, moving, accessing assistance, settling, and rebuilding their lives. While foundational to understanding forced displacement, territory is a concept that has been largely overlooked. When territory is factored into studies on forced displacement, it is nearly always associated with the scale of the state, as the examples above of some European states securing their territorial borders demonstrate. This linking of territory to the state is not unique to studies of forced displacement. Indeed, it is commonplace in public and political discourses, as well as in academic fields such as international relations, political geography, geopolitics, and political science to associate territory with states. Examining territory at the scale of the state is important to understanding forced displacement, yet throughout this book I show that there are other forms and scales of territory that are also vital to examining the many policies, practices, and experiences of displacement.

While the concept of territory is my primary lens for examining forced displacement, I also focus on refugees' *experiences* of displacement. Research and writing that focuses on the lives and experiences of refugees is growing, but it is still largely excluded from mainstream accounts and historical records (Marfleet 2016). Thus, there is a pressing need to learn more about the opinions, experiences, perceptions, and lives of displaced people (Chatty 2016). The statistics on numbers of refugees, like those that I began this book with, help to provide broad context about forced displacement, but such statistics can also dehumanize refugees, turning them into mere body counts (Hyndman 2007).[3] Humanizing refugees, on the other hand, highlights their complexity as people who are resilient and capable, even while struggling with trauma and loss (Culcasi 2019; Aleinikoff 1995). This, in turn, can also help to nuance normative discourses on displacement and refugees, including the particularly destructive discourse that refugees are "threats" to a state's territory, citizens, or economy. Such humanizing, nuanced views are particularly important for readers in the Global North, who may live in places where refugees are constructed as threats, and who may not have had many direct experiences with refugees.

The ways in which I approach my study of displacement, territory, and refugees' experiences come from a combination of theories and praxis within feminist geopolitics, critical geopolitics, and postcolonial and decolonial studies, all of which have contributed to the broad, interdisciplinary field of refugee studies (as well as more focused areas of camp, mobility, and borders studies). As I discuss in detail in the next chapter, these areas of scholarship are divergent in many respects, but they all share the goal of highlighting critiques, histories, discourses, and experiences that have been marginalized or suppressed. Feminist geopolitics and decolonial studies are generally concerned with the experiences of marginalized people, to humanize them, and bring their lives and knowledges to the forefront of public discussion. Critical geopolitics and postcolonial studies are particularly keen to critique dominant power structures and discourses, often with a focus on the Global North and imperial powers. Each of these four areas of scholarship has addressed the concept of territory to some degree, but it has been within critical geopolitics that territory has been theorized most deeply and critically. Yet the arguments and approaches to territory that exist within critical geopolitics have not been widely applied to refugee studies. Considering the formative role that territory has in policies, practices, and experiences of forced displacement, I hope that this book will help to fold complex conceptualizations of territory into refugee studies.

In the remainder of this chapter, I first provide an introduction to Jordan's complex displacement policies and practices, which stand in contrast to many Global North states. Second, I describe the numerous, crucial linkages between territory and displacement. Then, I introduce two specific ways that I approach my examination of territory and displacement, namely, senses of belonging and territorial imaginings. Last, I provide an overview of my research methods and reflect critically on my research process.

Displacement in Jordan

Since its independence from the United Kingdom in 1946, the Hashemite Kingdom of Jordan has had a sustained and significant history of accepting refugees. Protracted conflict over the territories of "Israel/Palestine" (a term I use to refer to the same territorial extent as the British Mandate of Palestine and Historic Palestine) since 1948, the 1991 Gulf War, the 2003 US-led invasion of Iraq, the 2011 Syrian war, and the rise of ISIS/Daesh in 2014 have all created mass forced displacements into Jordan. As a result, today Jordan has a remarkably diverse refugee population and, again, has the second-highest refugee population per capita in the world (International Labour Organization 2015; Saliba 2016; McCarthy 2017). With a total population of just over ten million people, approximately half of Jordan's residents are Palestinian refugees (Ryan 2011). Jordan also hosts 664,414 registered Syrian refugees and tens of thousands of refugees from other states (Chatty 2010a; Gorman and Kasbarian 2015; UNHCR 2022). During its seventy-six years as an independent state and a major refugee "hosting" state, Jordan has created, altered, and implemented starkly different policies and practices to manage the diverse refugee populations within its state-territory. These policies and practices range from granting Jordanian citizenship to Palestinian refugees from the 1948 war to building a massive refugee camp near the Syrian border in July 2012 and forcibly encamping some Syrian refugees (Massad 2001; Hanafi 2014). Notably, while Jordan has this significant history of accepting and including refugees within its borders, the state is not party to the major international laws and agreements on asylum seekers and refugees.

Jordan is a postcolonial state whose borders were drawn at the will of British and French powers at the end of World War I. Prior to its imperial creation, the territory that would become Jordan was part of the Ottoman Empire from 1299 to 1922. Under the Ottomans, the land that is Jordan today, as well as large expanses of land to the north, south, east, and west, had a large

Arab majority who had many historical, cultural, political, and economic interconnections and similarities. In many ways, these interconnections have weakened since the mid-twentieth century with the creation of independent states across the region, yet many remain and affect both the Jordanian government's policies toward refugees and refugees' experiences with displacement. The legacy of cross-border Arab connections, however, coexists with policies and practices that are entirely state based. Indeed, the idea of the independent territorial state is central to the Jordanian government's refugee policies and practices.

The millions of different Syrian and Palestinian refugees living in Jordan have all had unique experiences and maintain varied senses of belonging to different territories. Some have found new homes in Jordan and consider themselves Jordanian. Many feel that Jordan is quite similar to the places from where they (or their families) were displaced and that they are thus in a "sister" territory among their "brethren." Others feel out of place and desire nothing more than to return home. This range of experiences and senses of belonging is not too surprising considering the different contexts of their displacement and different lengths of time spent in Jordan. Yet despite these important differences, refugees in Jordan share many commonalities. Most broadly, these refugees (or their immediate ancestors) were forcibly displaced over imperially imposed borders that have existed for just one hundred years, they were all granted refuge in Jordan, and they have all remade their lives there.

Territory and Displacement

As noted above, there are many relevant factors and concepts to consider when examining forced displacement. My focus on territory is not meant to dilute other important factors but to address the understudied ways that territories configure into forced displacement. As I elaborate throughout this book, different forms and scales of territory are foundational to the international refugee regime (IRR), state policies, and the experiences of refugees.

I have referred to the concept of territory many times above, usually in relation to the form and scale of the state (e.g., Global North states securing their territorial borders). Associating territory with states is the most common and dominant way that territory is conceptualized. There is no singular definition of territory, but Google Dictionary—arguably a widely read and influential reference—defines territory as "an area of land under the jurisdiction of a ruler or state." Similarly, academic definitions commonly refer to the modern concept of territory as "the spatial extent of state power" (Dahlman

2009, 77). The bonding of territory to states—what I refer to as the "state-territory nexus"—is not just textual or discursive but also material. This is clearly evident in the fact that for more than a century the state-territory nexus has been *the* foundation for the geopolitical division of the world. The state-territory nexus, as Steinberg (2009, 469) articulates, is our "universal geopolitical reality." Globalization and transnationalism have questioned and weakened this nexus in many instances (e.g., the EU), but the "universal geopolitical reality" of state-territories continues to dominate our political organization of the world into clearly divided, discrete territorial units, and this nexus is, in many ways, strengthening. Indeed, as noted above, we are currently witnessing new and intensified territorial-state securitization discourses and practices in many Global North states as they attempt to deter displaced people from seeking asylum at their territorial borders. The state-territory nexus is also foundational for international refugee laws and the broader IRR (see chapter 2 for details on the IRR). For example, for a displaced person to have legal status as a refugee, they must have crossed an international territorial-state border. The "durable solutions," which are official policies of international refugee organizations, seek to restore permanency of each displaced person as a member in one territorial state. In other words, the durable solutions work pointedly toward ending refugee status by fixing displaced people into state-territories where they can (theoretically) belong. While the state-territory nexus dominates the political ordering of the world, frames major international laws about refugees, and has countless effects on where refugees move and settle, there are other less-acknowledged forms of territory that have little to do with the state but are nevertheless important to examining forced displacement.

Nonstate forms and scales of territory that intersect with the IRR and refugees' lives are quite numerous. For example, detention centers, checkpoints, border crossings, berms, and militarized zones are forms of territory that materialize from the restrictive practices of managing forcibly displaced people. These forms of territory are often *associated* with a state or overseen by a state government, but they are not bonded in a nexus with the state. There are other nonstate-territories that have substantial impacts on forced displacements but are more fully decoupled from states. These are often less recognized in public and academic discourses, largely because they do not fit into the dominant political ordering of the world. As I explicate in chapters 4, 5, and 7, pre-imperial, anti-imperial, transnational, and hybrid territories, as well as refugee camps, all maintain territorial characteristics but are not bound to the state. As the title of this book implies, and as I elaborate in chapter 1, I seek to "displace" the conventional state-centered definition of territory (i.e., the state-territory nexus) and consider other scales and forms of territories that affect refugees' senses of belonging, their decisions about moving and settling, and the formal policies and practices of both the IRR and Jordan.[4]

Definitions of territory have changed and will continue to do so. Thus, my framing of territory is not universal or final. Unlike the conventional understanding (i.e., an area of land under the jurisdiction of a ruler or state), my definition of territory centers the power-laden processes of spatial inclusion and exclusion. At the broadest level, I define territory as *an ordering of space that is formed through the politics of who and what belongs. This definition then logically also means that territory is formed through the politics of who and what are excluded or who or what would need an invitation or agreement to be included.* My definition and framing of territory has evolved directly from my research on forced displacement, in which inclusion and exclusion within different territories is hugely impactful on policies and practices surrounding displacement, as well as on the lives of refugees. Nevertheless, I believe that focusing on inclusion/exclusion could also be relevant for elucidating insights into other territory-related topics, such as land conservation, urban segregation, and border conflict.

Senses of Belonging

The idea of "belonging" may seem mundane and self-evident, but it is a complex social and political concept (Yuval-Davis 2006). In general, "belonging" is an act or feeling of inclusion and connection with other people, groups, or things. People often feel a sense of belonging based on social groupings like class, race, gender, or generation; or perhaps by political party, taste in music, or religious affiliation. Also, as I examine in this book, people commonly attribute parts of their senses of belonging to different geographic places or territories.

Feelings or acts of belonging are to some degree always power-laden, *political* processes (Wright 2015; Antonsich 2010). Belonging is inherently political because the process of inclusion is simultaneously one of exclusion. In other words, for someone (or something) to be included means that other people (or things) must be excluded.

Belonging is also a *social* process. It is not simply an *individual's* feelings of inclusion or connection. Senses of belonging are formed through intersecting collective discourses, laws, and practices. Therefore, as Antonsich (2010) notes, to examine belonging means considering individual feelings and experiences, but always within the broader social and political contexts in which belonging manifests.

Geographic belonging can develop at a range of scales, from one's hometown, to a city, to the entire earth. Senses of belonging to a territory are similar but generally expressed in more pointedly political ways, in which strong feelings of inclusion/exclusion are inherent to the sense of belonging. For example, territorial belonging is often expressed as a person's lawful inclusion in a state where they have legal citizenship.[5] Yet, it is important to decouple territory from the state and to consider the other ways that inclusion/exclusion manifests geographically. As I discuss in subsequent chapters, nonstate-territories like cross-border historical regions or refugee camps are greatly defined by who is included and excluded.

Displaced people, like most people, have multiple senses of belonging to multiple places and territories (Marcu 2014; Marlowe 2017). However, unlike people who have not had to flee, displaced people's territorial senses of belonging are often disrupted or reconfigured in dramatic ways. Feelings of connection to the places and territories from where they fled are often heightened. Memories and imaginings of the places and territories where they feel they belong can be intense and evoke strong emotions of longing to return. At the same time, new connections and senses of belonging also commonly form as refugees migrate, settle, and rebuild their lives. Crucially, senses of belonging can directly affect a person's willingness to flee, where they flee to, and how they adjust and cope if they are given refuge; and thus senses of belonging are an important factor for examining forced displacement.

Territorial Imaginings

Research on "imaginings" has been a part of geographic scholarship for decades and has been the focus of several seminal pieces of work. For example, Doreen Massey (2005) uses the term "spatial imaginaries" to refer to the ways that hegemonic powers imagine space. Likewise, Edward Said (1978) uses the term "geographical imaginings" to denote the discursive ways in which Western, imperialist powers divided the world. Massey and Said both focus on how such imaginings justified imperialist actions. More specifically, Massey (2005, 4) argues that Cortez crossed the Atlantic to conquer Tenochtitlan in part because the Spanish Empire's Eurocentric imagination relegated the Aztecs to the status of lesser people. Said's seminal argument is that European Orientalists imagined the Orient as inferior and Orientals as backward. Such imaginings elevated Europeans as superior and served as a major justification of centuries of European imperial domination, as well as more recent neo-imperial US geopolitical actions like the 2003 invasion of Iraq.

While geographic imaginings at the level of dominant powers have been fairly well studied and critiqued, examinations into territorial imaginings of marginalized people and individuals are not as widespread (Glăveanu and Zittoun 2017, 3). It is, however, valuable to study imaginings of marginalized people because their geographical imaginings and senses of belonging often differ from mainstream conceptualizations (Jeffrey 2020, 34). Moreover, their imaginings can have direct and powerful impacts on their lives and actions. As Wendy Wolford (2004) demonstrates in her examination of marginalized people in Brazil, their spatial imaginaries have poignantly affected their actions and engagement with social movements fighting for land rights.

In this book, I address imaginings of territory that have informed displacement policies and practices within the IRR, most notably the state-territory nexus. I also examine dominant territorial imaginings stemming from Jordanian leadership that have affected their policies toward displacement. Some of these territorial imaginings have challenged imperial territorial divisions and borders, while others have reified them. Shifting from these rather macro-scale, hegemonic imaginings, I also narrow to focus on the territorial imaginings of displaced Palestinian and Syrian refugees in Jordan. As detailed in chapters 4–7, there are many forms of territory that they imagine and experience, some of which challenge the state-territorial nexus and others of which recreate it. Crucially, their imaginings do not merely circulate in their minds but, as I will detail in the later chapters, also have significant effects on many of their experiences as refugees in Jordan, including their decisions to move and settle and their senses of belonging.

Research Methods

My research and data collection methods are multifaceted. I draw from official documents and reports from the United Nations High Commissioner for Refugees (UNHCR), the United Nations Relief and Works Agency for Palestine Refugees in the Near East (UNRWA), and several other international nongovernmental organizations (INGOs), as well as key Arab League and Jordanian laws. I was also able to conduct several expert interviews with officials in different refugee aid organizations in Jordan. These two forms of data are the foundation of my discussions of the IRR and the centrality of territory within that complex regime. I also conducted fieldwork observations from within refugee communities in Jordan and 126 interviews with refugees and refugee aid workers. These two methods allowed me to study the experiences of refugees and to tease out the complex ways that territory figures in their lives.

I took six research-related trips to Jordan between spring 2011 and spring 2018. The trips ranged from one week to six weeks long. I conducted formal and informal interviews in many different places in the cities of Irbid and Amman; in the Palestinian refugee camps of al-Husn, Marka/Hitten, and Jerash; and in the Syrian refugee camp Za'atari.[6] My interviews with Syrians and Palestinians included specific questions about how they imagine the different state-territories that they have moved between. I also asked them about connections to their homes and homelands, to the Arab world, and to camps and communities within Jordan. I asked about their senses of belonging, their feelings of being included and excluded in Jordan, their preferred places to live, and where they call home. There is no direct translation for "territory" in Arabic. But other words like *'ard* ("land"), *watan* ("homeland"), *balad* ("country"), *dawla* ("state"), and *qutr* ("region" or "state") were commonly used in ways that echo my definition of territory as spaces of inclusion/exclusion. That Arabic lacks a direct translation for "territory" demonstrates quite clearly that the concept of territory that is so commonplace in the Global North is not universal.

I collected demographic data during interviews, but I do not provide comparisons between common social categories—age, gender, generation, citizenship, place of origin—as this is not a study about Palestinians compared to Syrians, nor a study of generations, citizenship, or gendered experiences.[7] While I recognize great value in such analyses, because they address heterogeneity and intersectionality, I did not design my project to examine differences between social categories. In the chapters that follow, I typically identify the age and gender of my respondents, not because of any causal significance of those two categories in determining their responses, but to add a humanizing element to my writing. I used Atlas.ti, a qualitative analysis program, to organize and analyze my interviews and field notes according to major themes, connections, and differences. I use terms like "often," "always," "commonly," "generally," "frequently," and "rarely" throughout this book to indicate commonalities and differences among my interviewees' answers, but I do not provide statistics or percentages about how they responded.

I examined the symbolic landscape in the camps, in homes, and in public spaces. I looked for signs and icons of nationalism, identity, and resistance, which are often displayed in graffiti, posters, flags, and maps. I spent considerable time collecting and analyzing maps and other cartographic representations from bookstores and the Royal Jordanian Geographical Center's library. National symbols, flags, maps, and other territorial representations are often symbolic parts of people's everyday territorial imaginings (Culcasi 2016). I often discussed common symbols and maps with my interviewees, as doing so invoked questions and discussions of places, states, territories, and belonging.

Conducting research and performing interviews has a clear power hierarchy. I was always keenly aware of the asymmetrical power between me—a white woman with US citizenship—and my interviewees (see Dempsey 2018). Nearly all the people I interviewed in Jordan—refugees and experts—welcomed me and expressed pleasure that I was engaging in this work. They wanted me to listen to them and, crucially, to share their experiences and stories with my students and wider audiences in order to educate people about their lives and help foster greater humanity. I made every attempt to be compassionate and respectful during the interviews and in my writing and presentation of their experiences and views. Great caution is needed in any attempt to "give voice" in general (Spivak 1988; Mohanty 2003) and for refugees specifically (Rajaram 2002). As literature in post- and decolonial studies and feminist geopolitics has asserted for decades, Western researchers like me risk projecting their own values and assumptions and thus improperly representing the stories of the people being interviewed. There is, as I will note again in the concluding chapter, also a danger in positioning myself as an "all-knowing" expert and belittling the knowledge of the people who shared their insights with me. Thus, while the words of the people I interviewed are necessarily filtered through my research process and training, I have made every attempt to share their complex imaginings in earnest.

Looking Ahead

In the next three chapters, I provide conceptual and contextual discussions of territory, displacement, and the intersection of the two in Jordan's policies and practices. In chapter 1, I bring together scholarship from critical geopolitics, feminist geopolitics, postcolonial studies, and decolonial studies in order to critique and advance conventional ideas of territory and the state-territory nexus. This chapter is largely a literature review of academic scholarship on territory. Readers who do not have a great interest in the history and theories of territory and are more interested in the topic of forced displacement may want to skim or skip chapter 1. Chapter 2 discusses the centrality and relevance of territory for analyzing the international refugee regime and displacement. In this chapter, I focus on the ways that territory intersects with displacement policies and practices, most notably the "durable solutions" of the IRR. Chapter 3 narrows the discussion to highlight Jordan's extraordinary refugee histories, geographies, policies, and practices. In this chapter I also provide a focused discussion about the Palestinian and Syrian refugee populations in Jordan. In chapters 4–7, I focus pointedly on the complex territorial imaginings and senses of belonging that Palestinian and Syrian refugees in Jordan maintain. In chapter 4, I discuss the ways that cross-border pre-imperial and anti-imperial territorial imaginings linger in the policies and practices of the Jordanian government and in the territorial imaginings and senses of belonging of Syrian and Palestinian refugees. Chapter 5 examines how pre-imperial and anti-imperial connections intersect with twentieth-century geopolitics and the state-territory nexus to create hybrid territorial imaginings, senses of belonging, and identities, which maintain elements of the state-territory nexus but do not conform to them entirely. Chapter 6 shifts to examine Palestinian and Syrian refugees' senses of belonging to the scale of the state and, more specifically, the state of Syria and the (quasi) state of Palestine. While the conventional state-territorial nexus is formative in the ways that Syria and Palestine are imagined, in this chapter I highlight that there are also coexisting ambiguous, amorphous, and abstract imaginings of these state-territories. I scale down in chapter 7 to examine refugee camps. Unlike the other forms of territory I examine, refugee camps exist only in relation to displacement. They are the result and materialization of the IRR's inability to find durable solutions and to better manage forced displacement. With seventeen official camps, Jordan has a complex "campscape" (Martin 2015).[8] In that chapter, I show that camps in Jordan have evolved into different forms of coexisting territories. A concluding chapter, chapter 8, summarizes my main findings, links these findings from Jordan back to the problematic discourse of "refugee crises," and makes some suggestions about ways to improve the IRR.

1

Displacing the Study of Territory

There have been many in-depth and ongoing studies of different spatial-political concepts like states, nations, nation-states, regions, and borders, but the concept of territory has become a serious and valued area of inquiry in Anglo-American social sciences and humanities in the past decade. There are a few different reasons for the neglect of this concept, including the racist, deleterious effects of Nazi territorial theories; decontextualized assumptions that territories are static, passive, and/or natural surfaces; as well as post–Cold War globalization that was purported to have weakened the power of states, nations, and territories. Yet, as many scholars have noted more recently, the concept of territory deserves greater attention and new theorizations because it is impactful and influential in countless aspects of life and politics (Jones and MacLeod 2004; Carter 2005; Newman 2006; Murphy 2013; Collyer and King 2015; Maier 2016; Elden 2009; Kolers 2009; Goddard 2010).

The study of territory was shunned by academics after World War II because of the racist and deadly effects that one particular territorial theory had in Europe. At the end of the nineteenth century, German geographer Fredrich Ratzel (1896) theorized territory as a biological, living organism that needed to grow. Decades later, the Nazi Party adopted Ratzel's theory of territory and developed the idea of *Lebensraum*, which purported that German territory needed to expand in order to exist. This territorial theory became a justification for the Nazis' aggressive territorial expansion and extermination of people who they believed could hinder the growth of the German nation. This sordid connection between a territorial theory and the horrors of WWII led to academics largely ignoring the concept of territory, which was particularly notable within the discipline of geography. Yet, tacit assumptions about territory continued to figure into some academic realms. Since the 1950s and continuing even today, territory has commonly been presumed to be a natural, static surface that humans and animals instinctively seek to control. The focus on territorial control had particularly significant impacts within international relations and mainstream geopolitics of the Cold War. Within both these discourses, territories were viewed as discrete entities that state powers owned and controlled. While the studies of human or animal behavior and international relations are quite distinct, both viewed territory as an uncomplicated, passive factor that did not need to be contextualized or theorized. Territory merely existed as a passive surface in which politics and life unfolds. In the early 1990s, the assumption that territory was a static, natural entity was recirculated. However, new critiques of territory, particularly within transnational studies, began to pointedly refute the idea that territory, along with states and nation-states, had much of a role or importance in understanding the post–Cold War era and globalization. As a result of increased human and capital mobility that surged in the mid- to late twentieth century, many scholars purported that economic, political, and cultural flows had usurped states, nations, and territory (e.g., Appadurai 1996b; Bhabha 1994; Field and Kapadia 2011; Monsutti 2010). The supposed borderless world that was evolving meant there was little reason to study territory.

Contrary to the views of territory outlined above, the field of critical geopolitics has argued that there is a great need to study the concept of territory. Crucially, critical geopolitics rejects the facile, decontextualized assumption that territory is a static, passive, or natural surface in which politics and life simply happen. Instead, it has asserted that territory is important in a globalized world and that it must be theorized in new ways that recognize its complex social and relational production. In the pages that follow, I review seminal pieces of academic literature within critical geopolitics that have revived the study of territory. Then, I discuss the ways that feminist geopolitics and postcolonial and decolonial theories are further advancing discourses on territory by focusing specifically on marginalized people and their imaginings and experiences of territory. My overarching goal in this chapter is to bring these bodies of literature together to create an approach to examining territory that underscores its social and relational production, which includes recognizing its evolving forms and scales and the existence of territories that do not necessarily fit the modern ordering of the world. But before delving into a review of academic literature on territory, I first provide a brief historical overview of the emergence of "modern" territory and the dominant system of dividing the world into discrete territorial states.

The Emergence of "Modern" Territory

The spatial organization of the world into discretely bounded territorial states—as we are accustomed to seeing on world political maps today—emerged from within Europe over hundreds of years.[1] During the Middle Ages, European territories, like territories everywhere else, were not necessarily discrete, state-like entities. Varied and overlapping systems of rule—such as feudalism, the church, and empires—had distinctive ideas and practices of control over land (Delaney 2005, 47; Sassen 2008, 31–71; Sack 1986, 92–126).

The "modern" system dividing the world into discrete territorial states is widely considered to have its origin in the peace treaties signed in Westphalia in 1648 (Sassen 2008, 53; Maier 2016). After decades of war, including the Eighty Years' and Thirty Years' Wars, leaders and delegates from across Europe signed several different treaties in hope of creating peace among their empires. These treaties established new forms of legal and political relations between European powers; or, as Osiander (2001, 270) states, they "confirmed and perfected . . . a system of mutual relations among autonomous political units." While this system of "mutual relations among autonomous political units" was a predecessor to our current ordering of the world into discrete territorial states, the treaties were not, contrary to popular belief, the "founding moment" of the world's current state-based international system (Elden 2013, 309). Instead, it took about four centuries, and many disparate events, factors, actors, and technologies, for the modern system of territorial states to evolve. For example, as Elden (2005, 2013a) has discussed in detail, advances in technologies of mapping, calculation, and surveying well after 1648 were key to producing knowledge about territories that could then be ordered and controlled. Another important event was the 1933 Montevideo Convention on the Rights and Duties of States, wherein states were defined formally as having four characteristics, one of which was a discrete, bounded territory (the three other traits were a permanent population, a government, and international recognition). Macro processes and doctrines of imperialism, industrialization, capitalism, and nationalism, particularly in the nineteenth and twentieth centuries, spread the idea of state-territories outside of Europe and disseminated the state-territorial system across the globe (Osiander 2001). In summary, while the state-territorial system evolved in Europe over hundreds of years (Painter 2010, 1094), for most of the rest of the world, the system of territorial states emerged in the early and mid-twentieth century as European empires fell and former colonies gained independence.

Critical Geopolitics and Territory as a Social Relation

As discussed above, studies on the concept of territory were uncommon from the mid- to late twentieth century. However, a few did exist, and those few offered new views on territory. Beginning with Jean Gottmann's 1973 book *The Significance of Territory*, a slow move toward critical studies of territory has evolved. His work was seminal in theorizing territory away from the perspective that territorial control is a natural part of human (and animal) behavior. Instead, Gottmann asserts that territory is a human and political process of organizing space into political jurisdictions (see also Ferraz de Oliveira 2021; Elden 2013b). Thirteen years after Gottmann's book, Robert Sack's (1986) book *Human Territoriality* developed the idea that humans created territory, likewise arguing that territoriality is not natural but the result of *social* processes.

Then, in 1994, John Agnew's article "The Territorial Trap" ushered in new critiques of territory that are still widely cited today. In his article, Agnew argues against the state-territory-centric thinking that had developed within international relations. He points to three assumptions about territory that were so commonplace in the 1990s that they had "trapped" our thinking. The assumptions are (1) that territory is equated with state power, (2) that territories have separate domestic and foreign political spheres, and (3) that territory is a container of politics. Agnew's critique of these unexamined assumptions about territory has been so influential that some academics have been hesitant to even engage with theories of territory because they worried that they might fall into this "trap" (Elden 2013a, 3). Gottmann's and Sack's seminal reframings of territorial theories to focus on humans' role in the production of territory, as well as Agnew's challenge of prevailing territorial assumptions within international relations, helped to launch many more investigations into the complexity of territory and its political and social production.

Stuart Elden has provided the most comprehensive and sustained writings on territory of any scholar. Building from these early critiques of territory, the argument Elden weaves throughout his books and articles—which span from textual analysis of ancient Greek writers to that of contemporary political speeches about the US-led war on terror and, most recently, Shakespearean ideas of territory—is that territory (as a word, concept, and practice) emerged historically and continues to evolve. Elden's work is seminal in highlighting not only that the idea of territory has changed but that it also has had many forms that are not linked to the state.

Swiss geographer Claude Raffestin (1984, 2007, 2012) is less well known in Anglo-American geography than the writers I noted above, yet his writings on territory are insightful and deserve greater attention (Klauser 2012). In 1984, he published an article titled "Territoriality" in the *International Political Science Review*. In this article, he argues that territory is both concrete and abstract. The "concrete," which he refers to as "geographic territory" and "spatial organization," is physical and morphological. The "abstract," which he refers to as "symbolic territory" and "social organization," is more cultural and social. The concrete and abstract, he argues, are not distinct but instead interface and complement each other. In this 1984 article (as in two subsequent ones in 2007 and 2012), he also argues that geographers must move away from examining the object or morphology of territory. Drawing from Foucauldian ideas of power relations, he posits the need to examine territory relationally and to understand how it comes into existence through social relations. For Raffestin—like Gottmann, Sack, Agnew, and Elden—territory is created and recreated through political and social relations as opposed to being a preexisting or natural entity (Murphy 2013).

Conceptualizing territory as a social relation is an ontological position that views territory as produced through and by people. While this position recognizes that material and geophysical dimensions matter in the production of territory, it is people and their social relations that matter the most. The emphasis on relations and production greatly echo Henri Lefebvre's (1991) seminal argument on the production of space. Lefebvre asserts that there are no preexisting, fixed, or inert spaces in which life just unfolds, and therefore we should examine not objects or things that happen in space but, instead, the ways that space is produced (see also Soja 1996; Massey 2005).

While the view of territory as being produced through social relations has slowly become more recognized in critical geopolitics, ideas of territories as discrete, fixed containers of life and politics are pervasive in academic discourse (public and political discourses too). Indeed, some scholars have posited that territory is incommensurable with relational thinking (Delaney and Rannila 2021). As Klauser observes, "territoriality and relationality have come to be seen as opposites" (2012, 110). Likewise, Painter (2008, 348) finds that many academics have continued to treat territory as qualitatively different from relational spatial organization. For example, Tim Cresswell, in his book *Geographic Thought* (2013), juxtaposes relational and territorial thinking, claiming the former focuses on "flows" and "things that lie beyond," whereas the latter focuses on "bounded entities within which particular things occur or are allowed" (73). This tendency to frame territories as bounded entities in which life simply occurs is still commonplace, but as noted above, there are now many different scholars who refute this position.

This ontological view of territory as being produced through social relations informs my work throughout this book. Focusing on the production of territory allows me to underscore the complexity and dynamism of territory (dell'Agnese 2013; Lefebvre 2009; Deleuze and Guattari 1987), including its multiple forms and scales and the ways that it evolves and changes. Further, as I discuss in the next section, I also draw on feminist geopolitics and postcolonial and decolonial scholarship in order to give pointed attention to marginalized people's role in production of territory, particularly through their imaginings of territory and their senses of belonging to territories. Again, I define territory as an ordering of space formed through politics of inclusion/exclusion, and therefore I focus on the political and geopolitical dimensions and relations of the production of different territories.

Feminist Geopolitics, Postcolonial Studies, Decolonial Studies, and Territory

Countless groups and people imagine, alter, experience, and produce territory. However, the views and experiences of groups and people in the upper echelons of power structures and knowledge production generally maintain disproportionate control over the production of territory. As a result, the views of marginalized people and their roles in the production of territory are often overlooked (Antonsich 2010; Delaney 2005, 34; Chakrabarty 2000). Likewise, views of territory that do not conform to dominant discourses and division have been largely subordinated (Sylvester 2012; Peters, Steinberg, and Stratford 2018; Quiquivix 2014; Sparke 1998). However, feminist geopolitics, postcolonial studies, and decolonial studies have each provided theories and tools to examine and highlight the significance of marginalized people's views, experiences, and lives in the production of territory.

There is great diversity in the themes and focuses of feminist geopolitics, but what unites this subdiscipline is the move away from the masculinized scale of the state and toward smaller scales that often include the home and the body.[2] Feminist geopolitics has not focused much on theories of territory (Coddington 2018a, 192), largely because the concept is viewed within the subdiscipline as a problematic reification of masculinist conceptions of the state, power, and geopolitics. Yet, in the past ten years, feminist geopolitics has begun to examine the concept of territory. At a broad level, a feminist approach to territory has challenged conventional concepts of territory as a bounded entity linked to the state and has helped highlight marginalized views that have been suppressed (see also Wastl-Walter and Staeheli 2004). There has been a particular focus on the ways bodies are gendered and territorialized as sites of power struggles.[3] For example, Sara Smith (2013) has examined the intersection of geopolitics and bodies through birthing and marriage to show how bodies both make territory and serve as a symbol of territory. Allison Hayes-Conroy (2018) examines how bodies and territories merge in resisting dominant power structures in a social movement in Colombia. And Sofia Zaragocin and Martina Caretta (2020) demonstrate the connections between bodies and territory in women's grassroot movements in Ecuador.

Postcolonial theories are also challenging dominant discourses and conventional views of territory by focusing on the ways that Western imperial and colonial pasts are a part of the present. These theories are pointed analyses and critiques of imperial geopolitics and state making. For example, postcolonial theories of territory assert that the "modern" state system of clearly defined and bounded states that we are so accustomed to seeing on political world maps today is more imperial or postcolonial than it is "modern" (Delaney 2005, 36). Recognizing the imperialist origins of state-territorial divisions today then logically also means that there are other, non-imperial territorial histories and imaginings that have been suppressed.

The study of decolonial geographies builds from feminist geopolitics and postcolonial studies, as well as indigenous geographies (Labrador and Ochoa 2019; Robertson, Okpakok, and Ljubicic, forthcoming) and critical race theory (Hawthorne 2019). It explicitly focuses on sociospatial relations *of people* who were once under colonial rule or are currently within settler colonial states. In other words, it refrains from the postcolonial focus of critiquing imperialist powers (Radcliffe 2017; Naylor et al. 2018) and instead highlights the ideas and actions of colonized people. For example, decolonial studies has focused on the ways that formerly colonized people across South America have mobilized to assert their rights and claims to territories that were taken from them by European colonists (Halvorsen 2019; Hayes-Conroy 2018; Ballvé 2012; Clare, Habermehl, and Mason-Deese 2018; Bryan 2012). Decolonial studies is, therefore, also commonly involved in the political praxis of reasserting land claims and regaining territories that were colonized.

In Summary and Looking Ahead

Defining and framing territories as discrete, bounded entities linked to the state, as natural or preexisting entities in which life unfolds, and as a concept that materialized directly from European peace treaties in the seventeenth century are all commonplace and mainstream ideas, but these are also "a fiction" (Billé 2020, 3). As Billé further argues, the cognate concepts of states and nation-states[4] may be fiction, but they are a "potent fiction . . . to which we are all committed and in which we remain deeply invested." In 2009, fifteen years after his critique of the "territorial trap," Agnew (2009, 29) found the concept of territory to be still "fatefully tied to the modern state." Likewise, Murphy (2013) reasserts that there is a "continuing allure of territory" based on conventional territorial ideas. He similarly observes that "the territorial norms of the modern state system display a remarkable tenacity" (1214). Indeed, the recent increase in border securitization by many Global North states, as discussed in the introductory chapter, demonstrates that conventional ideas of territory and its nexus with the state remain utterly formative and impactful today.

Yet, as I have discussed in this chapter, there are several areas of scholarship that have provided critiques of the conventional, Western understanding of modern territory. These include positioning territories as being produced through social relation, as opposed to seeing territories as bounded, discrete containers in which life or politics happens. Examining the many dimensions involved in the production of territories shows that territories are always being made and thus are not static, natural, or preexisting. Feminist geopolitics, postcolonial theory, and decolonial scholarship likewise provide both critiques of conventional territory. Crucially, these areas of scholarship all recognize and highlight marginalized, non-Western, and non-imperial conceptions of territory, and they are cultivating nuanced and critical ways to move the study of territory forward.

Throughout the remainder of this book, I examine the many ways that the "modern" ordering of the world into discrete territorial states remains central to the practices and experiences of forced displacement, but I do so by underscoring how territories are produced, dynamic, and evolving. Further, I also show that other forms and scales of territory, which are non-Western and/or decoupled from states, also exist and have formative impacts on forced displacement and vice versa.

2

The International Refugee Regime, Durable Solutions, and Territory

The political ordering of the world into state-territories, what is commonly referred to as the "modern state system," has immense effects on the formation of policies and practices related to migration and forced displacement. For example, the legal definition of a refugee and many of the international refugee regime's practices and policies are based on the normalization of the state-territory nexus. The legal definition of a refugee, as in the 1951 Convention Relating to the Status of Refugees, is a person who leaves their place of origin and crosses an international border into another territorial state to seek protection from persecution (Betts 2013, 74). Thus, as Emma Haddad (2003, 297) states quite simply, "without the modern state there could be no refugees."

Further, the foundational international refugee laws position state-territories as containers where people belong as sedentary residents or citizens.[1] Sophia Hoffmann (2016a, 344) has argued that international refugee laws and protections are based on the idea that "every human being is naturally linked to a sovereign state." The state-territory nexus and the ideal of people belonging to one state is also clearly demonstrated through the "durable solutions" of the UNHCR. Each of the durable solutions pointedly seeks to stop refugees from moving across international borders by reterritorializing them as sedentary people linked to a single state-territory. The goal of sedentarizing displaced people has some practical rationales, like providing opportunities for displaced people to resettle, but in the process it also categorizes refugees as people who are aberrant, as deviating from the norm of belonging to one state, and even as potential threats to states because of their statelessness (Malkki 1992; Zaman 2016; Watkins 2020). Moreover, the protracted status of millions of refugees for whom no "solution" has been found demonstrates that the ideal of sedentary citizenry living within one state-territory has not been realized.

In this chapter, I illustrate the centrality of the modern state system for the policies and practices of the IRR and displacement more broadly. Focusing on formal and official territorial discourses and laws, I show the different ways in which the state-territory nexus is foundational to the guiding principles and practices of the IRR. The first subsection below is an introduction to the IRR, the UNHCR, and the formative legal documents of the regime. In the second subsection, I discuss common labels and categories used within the IRR, noting confusion and debates over the terms "refugee" and "migrant." The regime's development and promotion of the "three durable solutions"—of repatriation, settlement in the place where refuge is first sought, and resettlement in a third state—is the focus of the third section. There, I discuss the ways in which these "solutions" are state-territory-centric and seek to reterritorialize and sedentarize displaced people. In the fourth section, I review some alternative approaches to refugee management that shift away from prioritizing the reterritorialization of refugees in one territorial state and instead work toward improving refugees' quality of life and protecting their human rights, including their right to mobility.

The International Refugee Regime

During the early and mid-twentieth century, Western powers created the IRR within the context of their own political interests and visions of the modern state system. The regime has since evolved into a massive, complex global system. There is no universal or singular regime, but countless actors, laws, policies, practices, and materialities that constitute it. The IRR includes organizations (humanitarian, development, labor, and security), state governments, aid workers, migrants, treaties, fences, detention centers, and so on. The regime, however, is anchored significantly by the United Nations Office of the High Commissioner for Refugees (UNHCR). The UNHCR, and the regime by extension, are, as I discuss below, "trapped" within the conventional territorial norms that Agnew and many others have critiqued.

The post-WWII era is generally considered to have been the most formative for the creation of international refugee laws and policies. In 1950, in the wake of WWII's mass displacements across Europe, the UNHCR was established, and the key international legal instrument for governing refugees was ratified soon afterward in 1951. There were, however, important antecedents stemming from the aftermath of WWI, including the creation of formal refugee organizations and of international policies and practices to manage forced displacement (Culcasi, Skop, and Gorman 2019).

The post-WWI years were a crucial time when the Western, Allied victors of the war promoted, disseminated, and normalized the state-territorial ideal and simultaneously cultivated what would become the predecessors of the UNHCR and the 1951 Convention Relating to the Status of Refugees. In 1921 Fridtjof Nansen, a Norwegian diplomat, was appointed by the League of Nations as the first ever High Commissioner for Refugees. Nansen's work focused first on repatriation and aid for Russian and Armenian refugees but his mandate soon expanded. In 1930, upon Nansen's death, the League of Nations established the Nansen International Office for Refugees, which soon became the most active organization for addressing refugee issues during the interwar years (Haddad 2003, 316). Some of the office's most significant work was the creation of the "Nansen passports," which were certificates that allowed stateless people to move across European borders (often for the purpose of obtaining work).[2] The Nansen office established a system of managing and documenting refugees at a scale never seen before. In 1938, in recognition of its leadership in refugee assistance, the Nansen office was awarded the Nobel Peace Prize.

The first international agreement on refugees is not the well-known 1951 Refugee Convention but the 1933 Convention Relating to the International Status of Refugees. Focused on WWI refugees in Europe, the 1933 convention created some of the world's most formative refugee laws (Jaeger 2001). For example, article 3 of the 1933 convention declared that a state cannot "remove or keep from its territory . . . refugees who have been authorized to reside there regularly." This article created the principle of non-refoulement, which remains a pillar of international refugee policies today.

In December 1949, after the unprecedented mass displacements of WWII, the newly formed UN General Assembly established the Office of the High Commissioner for Refugees (UNHCR). The second general principle of the UNHCR Statue, adopted in December 1950, asserts that the commission's work shall be nonpolitical, and instead humanitarian in scope.[3] The statute outlines the functions of the Office of the High Commissioner and calls on state governments to cooperate with the UNHCR. The first "general provision" of the statute declares that the UNHCR will seek "permanent solutions for the problem of refugees." These "solutions" are described as "the voluntary repatriation of such refugees, or their assimilation within new national communities." In seeking either of these "permanent solutions for the problems of refugees," the UNHCR's core mandate is as much a humanitarian one as it is a state-territorial one, as the solutions attempt to rectify refugees' aberrant status by reterritorializing them as sedentary people within one state-territory.

In July 1951, the Convention Relating to the Status of Refugees was adopted. At its base was a moral obligation to manage the mass displacement in Europe caused by WWII. This convention remains the most formative legal instrument in international refugee law today. It is a legally binding document that provides guidelines and obligations for its signatory states. The core principle of non-refoulement, stated in 1933, was reasserted in article 33 of the 1951 convention:

> No Contracting State shall expel or return ("refouler") a refugee in any manner whatsoever to the frontiers of *territories* where his life or freedom would be threatened on account of his race, religion, nationality, membership of a particular social group or political opinion [emphasis mine].

The legal definition of a "refugee" is from article 1 of the convention:

> As a result of events occurring before 1 January 1951 and owing to well-founded fear of being persecuted for reasons of race, religion, nationality, membership of a particular social group or political opinion, is *outside the country of his nationality* and is unable or, owing to such fear, is unwilling to avail himself of the protection of that country; or who, not having a nationality and being outside the country of his former habitual residence as a result of such events, is unable or, owing to such fear, is unwilling to return to it [emphasis mine].

The term "country" is used in article 1 and an additional forty-five times throughout the convention; the term "territory" is used in article 33 and another forty-seven times in the document. Throughout the convention, the terms "territory/territories" and "country" are used slightly differently but always with a connotation of "states." For example, article 26 pronounces, "Each Contracting State shall accord to refugees lawfully *in its territory* the right to choose their place of residence to move freely within its *territory* . . ." [emphasis mine]. The usage of the term "territory" in this article signifies the spatiality of the state. As many other articles of the convention detail (e.g., articles 4, 17, and 24), territory is a space in which the state is obligated to uphold the convention-secured rights of refugees. The term "country" is used in a similar, but not identical, way. "Country" only *implies* a connection with a state, while "territory" is used in *direct* connection to the state. For example, the possessive "its" is used seven different times to refer to the relationship between territory and the state, whereas "its" is not used to refer to the relationship between country and state. This is a subtle distinction, yet one that demonstrates that the state-territory nexus is normalized and reified within the discourses of the convention (and the wider IRR). Whether the legal definition of a refugee or the "solutions for the problems of refugees," the international refugee regime's foundations and goals are decidedly state-territory-based.

There is a tension, if not a contradiction, between the convention's mandate to protect refugees and other twentieth-century international laws and principles. International refugee laws prioritize the protection of *refugees*, whereas many other international principles—including those stated within the UN Charter—guarantee a state's right to protect its *sovereignty*, *territory*, and *borders*. The right for a state to protect its territory has been leveraged repeatedly as a justification to deny asylum seekers and refugees entry and basic protections. As I discussed in the introductory chapter, many states have securitized their borders and increased bordering practices to stop displaced people from crossing into their state-territory and obtaining protection (Hoffmann 2017). The US, Australia, Italy, and Hungary (and others) are signatories to the international laws governing the protection of refugees, but through their exclusionary and restrictive practices they are violating the laws they signed, as well as disregarding the overarching humanitarian morals that underwrite the laws. In other words, such states are defaulting on their obligations to help refugees, in preference of implementing restrictive policies that hinder asylum seekers' and refugees' ability to access their territories and obtain protection if they are able to enter (Bloch 2020, 437).

The 1951 convention had clear geographic and temporal rules that further defined who could be a refugee: a refugee was someone within Europe who had been displaced as a result of WWII. Yet as decolonization and independence movements gained momentum across the globe and as the Cold War exploded in the aftermath of WWII,[4] the refugee situation of WWI and WWII were no longer unprecedented or unique. To address the new and generally non-European refugee populations, the 1967 Protocol Relating to the Status of Refugees was adopted in order to remove the 1951 convention's geographic and temporal restrictions. The same principles and foundations were retained, but its scope was now global.

A total of 148 different states are party to the 1951 convention or the 1967 protocol (or both). Thus, there are about 50 nonsignatory states globally. There is a common assumption that signatory states are "superior" to nonsignatory ones because they are, theoretically, more open and provide better services to refugees. But as with most assumptions, there is value in critiquing it. As Coddington (2018b) argues, signatory and nonsignatory states are not necessarily different in how they apply and practice refugee management, and signatory states like the UK, she notes, are increasingly providing subpar services and aid levels that parallel those of nonsignatory states. Indeed, the quality of protection is declining in some signatory states, and some nonsignatory states are providing higher standards of aid. As I discuss in chapter 3, Jordan is a nonsignatory state that hosts millions of refugees. And while its management is varied and far from perfect, it has provided protections and opportunities for millions of refugees for decades.

Labels and Categories (and the Problem with Them)

The terms "refugee," "displaced person," "asylum seeker," and "migrant" are at times used synonymously, or they can be used to denote different categories of people (Ehrkamp 2017a; Hyndman 2009). In general parlance, a refugee is someone who has been forcibly displaced from the state where they reside and has crossed international borders. As noted above, the legal definition of a refugee is someone who is "outside the country of his nationality" due to persecution. Obtaining legal refugee status from the UNHCR means that the refugee is eligible for UNHCR humanitarian aid and is protected against refoulement. If a legally defined "refugee" is given permanent residency or citizenship in a new state, or if they repatriate back home, then generally they are no longer considered a refugee. "Asylum seekers" are people who have fled across borders in search of protection but whose legal status as a refugee has not yet been determined. In other words, they are awaiting refugee status. In many Global North states, these people are also commonly referred to as "irregular migrants," which indicates that their arrival in a new state was not planned or preauthorized by that state. Some asylum seekers will achieve legal refugee status in the state where they have sought protections, but many will not. The term "forcibly displaced person" (sometimes shortened to "displaced person") refers more broadly to a person who was forced from their home and is residing in a different state but does not necessarily meet UNHCR criteria or chooses not to seek legal refugee status with the UNHCR. "Internally displaced person" (IDP) refers to someone who is displaced from their home but has not crossed state borders. All these labels and categories are common within public and academic discourses on migration and displacement, as well as within the IRR, but at times there is confusion over these terms and categories. As I discuss in the next chapter, the Jordanian government uses different official labels to refer to people who are all ostensibly and legally "refugees."

There is some debate about how these terms relate to the label "migrant." People who move for economic reasons are often considered to be *voluntary* migrants, whereas refugees have been *forced* to move as a result of war or persecution. Again, when a person is legally a "refugee," the international community has a lawful obligation to protect that person, but this obligation does not apply to "migrants." Yet this distinction is not nearly as clear-cut as it may seem. Indeed, the overlap of economically motivated "migrants" and politically determined "refugees" is substantial. For example, conflict can lead to loss of livelihoods, economic stagnation, and poverty. The lack of economic opportunities and poverty can lead to crime, social instability, and conflict. These interconnected economic and political conditions—which can also include environmental degradation and climate change—can create the impetus for people to migrate in search of safety and well-being. Thus, the economic and political conditions that lead to people fleeing their homes often coexist, and thus the labels of "migrant" and "refugee" are not necessarily distinct (Bloch 2020, 74).

The Durable (Territorial) Solutions

The 1950 Statute of the Office of the United Nations High Commissioner mandates that the Office of the High Commissioner not only assist and protect refugees but also find durable and permanent "solutions" to the problem of displacement. In the 1950 statute, two such durable solutions are articulated: "voluntary repatriation" and "assimilation within new national communities." The 1951 convention reaffirms this goal of securing permanent, durable solutions for refugees. Since this mandate, the UNHCR and the IRR have advanced and slightly nuanced these two solutions, creating what are commonly known as the "three durable solutions," which are voluntary repatriation, integration into the host state, and resettlement in a third country. Today, seventy years since their inception, the durable solutions remain a "core mandate" and mission of the UNHCR (UNHCR 2015) and were included as a goal in the Global Refugee Compact of 2018 (Bloch 2020, 446).

Each of the three "solutions" seeks to formally end displacement and the aberrant status of refugees by reterritorializing displaced persons within a state-territory where they would legally belong. In other words, each of the durable solutions seeks to turn a refugee into a sedentary resident or citizen of one state. These three solutions have proven to be reasonable and durable in some situations, but they have not solved displacement or been a pragmatic solution for millions of refugees. Displacement is, in theory, a temporary situation that ends with resettlement or repatriation. However, 78 percent of refugees under the UNHCR's mandate remain in "protracted" displacement (defined as five years or more), which means none of the durable solutions have been obtained for these tens of millions of refugees.[5]

Repatriation is unequivocally the solution preferred by the UNHCR, individual states, and many refugees (Long 2013; Van Hear 2003).[6] Long (2013, 138), in her comprehensive analysis of repatriation, argues that this durable solution confirms "the 'rightness' of the national-state order by putting refugees back in 'their' place." But protracted conflicts, postconflict tensions, and the economic devastation that frequently coincides with conflict make returning home unfeasible to millions of refugees. Repatriation, as Long puts it, is a "seductively" neat solution and one that rarely works. In 2018, less than 3 percent of refugees voluntarily returned to their country of origin.[7] Integration into the host state where asylum was sought—the second solution—occurs, but not widely. Over the past ten years, 1.1 million refugees have been integrated into the states where they fled, meaning that they became citizens or permanent residents of the country that granted them asylum.[8] If we assume an even distribution of the second solution over the past ten years, then 110,000 refugees have been integrated into host states each year. Considering that there are currently around 70.8 million forcibly displaced people, 1.1 million over ten years is a minuscule percentage of people. Resettlement to a third country is also rare. In 2017, less than 1 percent of UNHCR registered refugees were resettled in a third country.[9] Further, this "solution" is increasingly untenable, as some Global North states have implemented increasingly restrictive resettlement measures that lowered the number of refugees they are willing to accept in a year. For example, in the 2020 US fiscal year, the Trump administration set its resettlement ceiling at 15,000 refugees, which is the lowest ceiling ever set in the US's forty years of having domestic refugee laws. While President Biden reset the ceiling to 62,500 in May 2021, this number will aid only a tiny percentage of the world's refugees.[10]

The difficulties in achieving the durable solutions are widely acknowledged by many actors, including the UNHCR and the tens of millions of refugees who remain in protracted displacement (UNHCR 2017; Van Hear 2003). Yet, regardless of the recognized failures, the durable solutions remain the preferred approach to refugee management globally and continue to frame international responses to "the problem of refugees." The durable solutions are greatly humanitarian and altruistic in intent, but as Bloch (2020, 453) argues forcefully, they continue to fail because they "are woefully inadequate in addressing the realities of displacement." Thus, there is an urgent need to recognize the shortcomings of the long-standing durable solutions, to reassess them, and to implement different strategies.

Alternatives to the Durable Solutions

Alternatives to the failing durable solutions already exist. There are many policies and practices facilitated through and between state and local governments, the UNHCR and other aid organizations, and from refugees themselves. In this subsection, I outline examples of policies that have offered alternatives to the state-territory-centric durable solutions.

There is diversity in the alternatives to the durable solutions, yet most approaches do not prioritize the goal of permanently reterritorializing refugees into one state-territory. Instead, they focus on the quality of life of refugees and on providing viable and humane options for refugees to rebuild their lives (Long 2013, 2014). Most alternatives to the durable solutions seek the attainment and protection of a variety of rights for refugees, including human, civil, political, residency, employment, and mobility rights. Generally, rights-based options will secure refugees' access to health care, education, and employment. Having "rights" also commonly includes the principle of non-refoulement, meaning that refugees have the legal right to remain in the state where they sought refuge and cannot be forcibly repatriated or deported (Kuch 2018).

Some alternative policies and practices specifically emphasize the right of refugees' *mobility* across state borders. This means that refugees are allowed to cross state borders freely for the purpose of employment or economic development. Mobility-centered approaches also allow displaced people to maintain different familial, social, and business ties across borders, whether these ties were established before, during, or after their displacement (Long 2013, 2014). Governments and INGOs have formally applied the mobility concept in several ways, most notably in facilitating family reunifications in different states and granting a variety of student and work visas that allow refugees to cross borders for education and employment.

Alternative approaches to the durable solutions are quite notable in the Global South, where different transnational agreements and organizations have provisions on migration and/or forced displacement (Van Hear 2003). These include the South African Development community, the Organization of African Unity, CA-4 in Central America, the Cartagena Declaration, the East African Community, the Kampala Declaration, and the Economic Community of West African States (ECOWAS).

Established in 1975, ECOWAS is one of the most established transnational organizations outside the European Union. In 1979, ECOWAS adopted the Free Movement of Persons, Residence and Establishment Protocol, which seeks to protect the free movement of people across the fifteen states that comprise ECOWAS. This protocol includes abolishing entry visas, issuing travel documents to ease movement, and granting refugees the rights to establish economic activities, to be employed, and to establish legal residency in a state other than their original one. Though this protocol has not been fully implemented, it has nevertheless supported local integration, economic development, residency, and visa-free movement of migrants, including refugees, for decades (Adepoju, Boulton, and Levin 2010). Implemented prior to the wars in West Africa in the 1980s and 1990s, this protocol has been applied to 117,000 Liberian and 18,000 Sierra Leonean refugees (Adepoju, Boulton, and Levin 2010, 137; Long 2013, 201, 206). While there are obstacles and challenges to ECOWAS's transnational, rights-based approach to refugee management, it does represent a viable alternative to the durable solutions.

There have also been attempts to modify and supplement the three durable solutions. For example, the official regional response plan for the Syrian war laid out by UNHCR, in consultation with other UN agencies and partners, and the countries of Iraq, Lebanon, Egypt, Turkey, and Jordan, builds from the three durable solutions while adding a fourth option. The official 2018–19 Regional Refugee and Resilience Plan (3RP) for the "Syrian crisis" asserts that refugees need access to durable solutions, in line with the core principles of international refugee law. The first three are the standard durable solutions, albeit with some nuance: (1) safe, dignified voluntary repatriation, (2) local solutions and opportunities, such as legal stay, and (3) resettlement to a third country. The first and third echo the original and keep within the goal of reterritorializing refugees, but the second implies temporariness. In other words, its intent is that refugees merely *stay* in host countries, as opposed to the second solution's original goal of permanent integration. In Jordan, there are complex explanations for this shift toward *temporary* stays, which I delve into in the following chapter, but briefly, this change is due to a general sense that host states are just too overwhelmed with the sheer number of refugees and do not have the resources to support refugees as citizens or permanent residents.

The fourth solution offered in the 3RP reflects the need for flexible options. This approach centers on offering *temporary* stay in a third country (as opposed to resettlement in a third country) through "humanitarian visas, family reunification, academic scholarships, private sponsorships, and labour mobility schemes." This fourth option seeks to temporarily resettle refugees in wealthy, UN convention–signatory states, but not to resettle refugees permanently in these third states. There are pros and cons for refugees gaining temporary refuge in a Global North state. It can benefit them by creating opportunities for them to become "self-reliant" and improve their quality of life. However, this "solution" (and the associated discourse of "self-reliance," which is commonplace in the IRR) can also have the effect of allowing Global North states to shirk the principle and obligation of offering permanent resettlement to refugees.

In Summary and Looking Ahead

In this chapter, I have highlighted numerous ways that the state-territory nexus lie at the foundation of the international refugee regime's policies and practices. The IRR's durable solutions and their commitment to reterritorialize refugees within one territorial state are, however, failing the tens of millions of refugees living in protracted displacement. Yet, concurrent with the durable solutions, there are examples of refugee management practices that are less state-territory-based. These alternatives generally focus on granting and protecting the rights of refugees and on allowing them to maintain and develop connections across state borders. The 3RP plan for Syrian refugees, which includes a fourth option that allows for temporary stays in signatory states, does not contradict the durable solutions or decenter the state-territory nexus but exists in tandem with them. As the next chapter highlights, the Jordanian state has numerous policies and practices of refugee governance that coexist to create a complex and diverse refugee system and landscape that both embraces and challenges the state-territory nexus.

3

Territory and Displacement in Jordan

Mass displacements and refugee crises are occurring across the globe, but they affect certain places more than others. The "Middle East" is generally identified as the world's biggest producer of refugees, while it simultaneously hosts a huge percentage of the world's refugees (Chatty 2010b, 277). Since the mid-twentieth century, conflict, war, and political instability have led to many different mass displacements across the "Middle East" (Gorman and Kasbarian 2015, 17).[1] The declaration of the state of Israel in 1948 led to the mass displacements of Palestinians regionally and globally. The 1979 Iranian Revolution, the 1980–88 Iran/Iraq War, the civil wars in Lebanon, Sudan, and Yemen, the decades-long conflict in Western Sahara, the American-led 1991 Gulf War, the 2003 US invasion of Iraq, the eleven years of war in Syria, and the rise of ISIS/Daesh in 2014 have each resulted in massive, forced displacements.[2] The huge majority of the refugees who have been displaced as a result of these conflicts have not attempted to reach European or other Global North states but instead have sought refuge in neighboring countries across the "Middle East."

I place the term "Middle East" in quotes in order to indicate that it is a highly problematic term. Briefly, the region that the term denotes is the only world region that lacks a continental referent, meaning that there is no particular continent relative to which the "Middle East" is situated. A quick look back to early twentieth-century history reveals that "Middle East" has its origins as a British term referring to the "middle" of Britain's "east." So this term that is commonplace throughout the Global North is an imperial-centric one (Culcasi 2010). Further, this term has not been widely adopted or used within the "Middle East." Indeed, it is often outrightly rejected due to its imperialist origins (Culcasi 2012). Less imperialist labels are increasingly common in scholarship and advocacy work that espouses decolonial and postcolonial views.[3] The acronym SWANA (Southwest Asia and North Africa) is slowly being adopted as one step in larger attempts to decolonize "Middle Eastern" studies. Thus, in the remainder of this book, I use the term "SWANA" to refer to a diverse and globally connected region that spans from Turkey southward through the Arabian Peninsula, westward across North Africa, and eastward to include Iran.

Situated geographically in the center of the SWANA region (figure 3.1), Jordan, unlike most of its neighbors, has not been a major producer of refugees. It has, however, hosted millions of refugees from neighboring states. Despite having scarce resources, high unemployment rates, and a struggling economy, as well as being a nonsignatory to both the 1951 UN refugee convention and the 1967 protocol, Jordan has kept its borders largely open and allowed the long-term residence and de facto integration of refugees, all while also providing basic humanitarian aid and social services like health care and education. Jordan's openness to refugees and its humanitarian support have not, however, been universal. Since the Syrian war, Jordan has implemented restrictive bordering practices—like border closures and forced encampment—to manage Syrian refugees both near and away from its border with Syria.

In this chapter, I focus on the Jordanian state's numerous policies and practices of refugee governance, highlighting the state's concomitant territorial openness and restrictiveness. First, I provide an overview of the creation of the state of Jordan in the twentieth century. Second, I review the cornerstone laws and policies that inform Jordan's refugee governance, highlighting some of the ambiguities and inconsistencies between policy and practice. Then, in two large sections, I discuss the mass displacements of Palestinians and Syrians into Jordan, focusing on the evolving policies, practices, and politics that affect their lives. Drawing from secondary literature, NGO reports, interviews with Syrian and Palestinian refugees, and interviews with experts on refugee politics in Jordan, this chapter illustrates that the Jordanian government has—at different times and with different displaced populations—mixed the seemingly contradictory practices of compassionate humanitarianism and territorial openness on the one hand with repressive securitization policies and border closures on the other. Ultimately, I show that there is no clear binary of Jordan being open *or* closed, nor of the state practices as being compassionate *or* repressive, but instead there is a complex and often disordered system of governance that attempts to manage the millions of refugees who have crossed Jordan's borders.

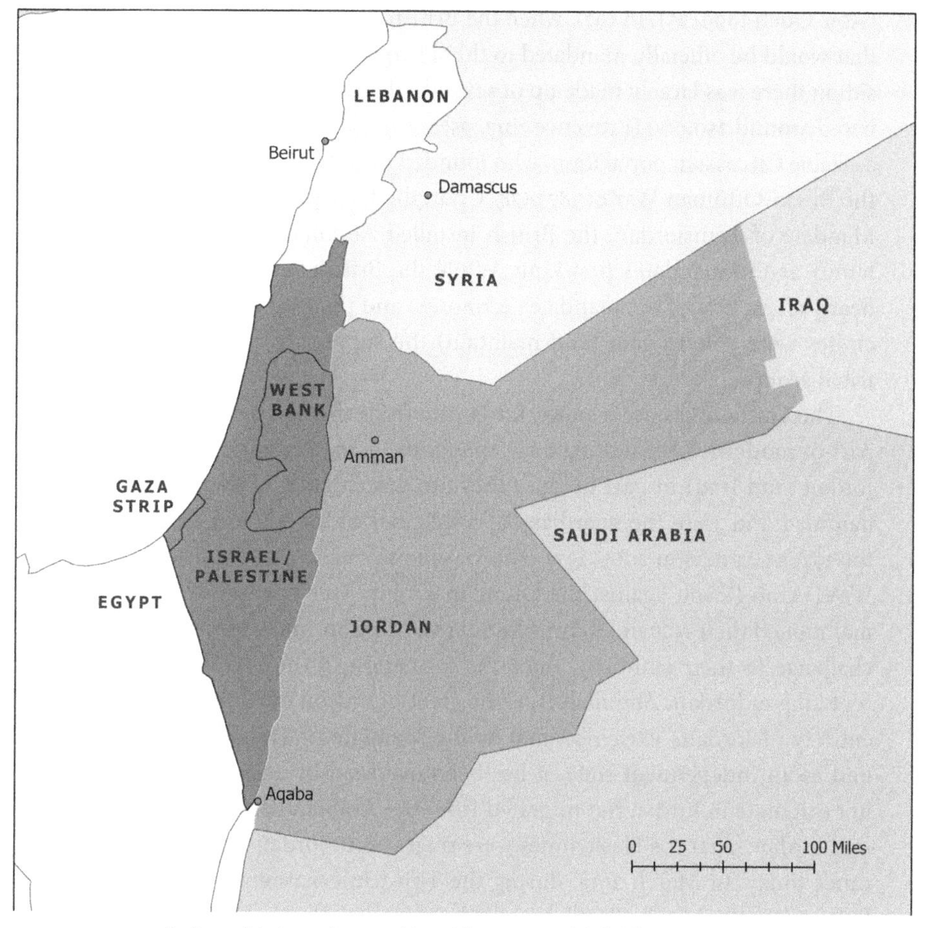

FIGURE 3.1. Jordan and its immediate neighbors. The territories labeled "Gaza Strip," "West Bank," and "Israel/Palestine" together comprise the area of Historic Palestine, which is the same region as the Mandate of Palestine. The Gaza Strip and the West Bank are also referred to as the "Occupied Palestinian Territories" because Israel has occupied them since 1967. I use the term "Israel/Palestine" to recognize the competing claims over this area, but it has been under the control of Israel since 1948 and is recognized by most international actors, including the Palestinian Liberation Organization, as Israel. The West Bank and Jordan were united under Jordanian authority from 1948 to 1967.

Producing the Jordanian State

Jordan has existed as an internationally recognized state for just under one hundred years. Parceled from the fallen Ottoman Empire at the end of WWI, the British Mandate of Transjordan was ratified by the League of Nations in 1923. The establishment of this imperial mandate set the foundation for the creation of the independent Jordanian territorial state in the 1940s

(Abu-Odeh 1999, 25). In 1921, when the British gained control of the territory that would be officially mandated to them two years later, the population residing there was largely made up of several different tribal groups who numbered around 250,000 (Provence 2017, 95; Schwedler 2022, 29). There was also a sizable Circassian population, who founded the city of Amman after fleeing the Russo-Ottoman War of 1877–78. Upon the formal creation of the British Mandate of Transjordan, the British installed Abdullah I of the Hashemite family as the mandate's first king. While the British maintained control of nearly all aspects of the mandate's economic and political spheres, the Hashemites were able to gain (and maintain) the support of many of disparate tribal groups.

The Hashemites are a powerful family from the Hejaz region, which is part of modern-day Saudi Arabia. The family gained legitimacy as rulers of Jordan (and Iraq) in part because they are descendants of the prophet Muhammad and were the guardians of holy cities of Mecca and Medina until the 1920s (Anderson 2005, 197). The Hashemites were also the leaders of the WWI Arab Revolt against the Ottoman Empire (see chapter 4). They have maintained their rule in Jordan without interruption, and with only very little challenge to their authority, since the formation of the mandate. The current king of Jordan, Abdullah II, is the great-grandson of Abdullah I. For the entirety of Jordan's existence, first as the Mandate of Transjordan and second as an independent state, it has been overseen by a monarchy that did not originate in Jordan but migrated from the Arabian Peninsula northward into Jordan. That the Hashemites were migrants to Jordan maintains significance today. In March 2018, during the Third International Conference for Refugees in the Middle East [*sic*], which I attended, Prince Hassan (grandson of Abdullah I and uncle of Abdullah II) spoke to the crowd and called for humanitarian compassion toward migrants and refugees. He reminded the audience that "people are people" and that his grandfather, Abdullah I, the first king of Jordan, had been born in Mecca and migrated with his family to Jordan. The history of Jordan, he stressed, is one of welcoming migrants.

The United Kingdom granted Jordan its independence in 1946, after twenty-four years of controlling the Transjordan mandate. The newly independent, postcolonial state was soon renamed the Hashemite Kingdom of Jordan. Over the next seven decades, the Hashemite monarchy created and implemented many state- and nation-building projects in order to unite the rather sparse population that lived within the imperially defined borders of Jordan (Culcasi 2016, 22; Anderson 2005).[4] Jordan had no united independence movement, nor a continuous past to recall. In other words, Jordan is not the "product of some slow evolutionary process" (Milton and Hinchcliffe 2009, xiv). Instead, it is the result of early twentieth-century imperial design and of mid-twentieth-century state- and nation-building initiatives from the monarchy, which, as Schwedler (2022) recently argued, were both conditioned by political revolts and protest that happened periodically across Jordan.

While the Hashemite family has been integral to the creation and ruling of Jordan (Anderson 2005), there are, of course, other factors in the construction of the Jordanian state and nation. A large majority of people residing in Jordan are Muslims (about 93 percent), and this has helped to create a sense of commonality and interconnections (El-Abed 2014, 85). There are also well-established familial and tribal alliances, Bedouin influences, and an ancient history that highlights the glorious Nabataean city of Petra. Arab nationalism, which I discuss in detail in chapter 4, has likewise played a key role in producing, defining, and uniting Jordan.

Jordan's Refugee Practices

Jordan has the second-highest ratio of refugees to citizens of any state in the world (neighboring Lebanon has the first) and the fifth-largest number of refugees in absolute terms (McCarthy 2017; International Labour Organization 2015; Saliba 2016; Francis 2015). Palestinians made up Jordan's first and largest wave of refugees. Syrians fleeing the civil war constitute Jordan's second-largest refugee population. The third-largest group is Iraqis, who were displaced as a result of the US-led 2003 invasion of Iraq and the conflicts that have extended through the rise of ISIS/Daesh in 2014. There are 66,823 registered Iraqi refugees in Jordan and an estimated 390,000 nonregistered Iraqi refugees (Bidinger et al. 2014). In addition to these three large populations, several smaller waves of refugees have entered Jordan since the 1970s. Refugees came (and some later left) from Syria in 1982 after a massacre in Hama sanctioned by the Assad regime; from Lebanon during its civil war in the 1970s; and from Yemen since war broke out there in 2014. Somali, Sudanese, Eritrean, and Chechen refugees, as well as Palestinian refugees from Syria (PRS), have also sought refuge in Jordan in the past two decades.

Whereas border controls, offshore deterrence, forced detention, and restrictive border policies have become commonplace in many Global North states, such practices are less common in Jordan. In many ways Jordan's policies and treatment of refugees are more progressive, generous, and welcoming than those of many wealthy Global North states (Francis 2015; Chatelard 2010b).[5] However, the rather generous and open practices that were the norm for decades in Jordan have been waning, particularly since the Syrian war (Chatty 2018).

Jordan's history of accommodating millions of refugees, many of whom remain in Jordan as protracted refugees, has taken place without any guiding laws that pertain specifically to refugees (International Labour Organization 2015, 23). Jordan is not a signatory to either the UN refugee convention or its protocol (see chapter 4 for an explanation). Instead, a mixture of different domestic laws and agreements have coalesced into a rather disordered, ad hoc, inconsistent, and evolving system of refugee management (Saliba 2016). There are some guidelines for managing refugees in Jordan, but these are not laws, and many of the guidelines are ignored or applied inconsistently, which creates quite a lot of confusion and ambiguity among state actors, aid organizations, researchers, and refugees.

The disordered system of managing refugees is evident, for example, in how the state categorizes different refugee populations. Of the many displaced people who live in Jordan, some are "refugees," others are "displaced persons," some are "asylum seekers," many are "guests," and millions have the labels of both "refugee" and "citizen." Palestinian refugees in Jordan have different legal status, identity cards, and rights depending on when they or their families arrived. Registered Syrian refugees in Jordan have legal status as refugees and are under the auspices of the UNHCR, but the government labels them as "guests" (*dyuf*) and asserts that they are temporary. Likewise, Iraqi refugees registered with the UNHCR are legally "refugees," but the Jordanian government labels them "guests" as well.

Jordan's varied, ad hoc, and evolving policies are not random or arbitrary, but quite strategic. Maintaining such policies can be quite beneficial as it allows decision makers flexibility and leeway in how they apply policies. And crucially, having unclear policies allows the state to modify practices to fit strategically within its evolving domestic politics and economy, its shifting national discourses, and changing regional geopolitics. For example, Jordan's refugee policies and practices have changed in stride with domestic tensions with the Palestinian Liberation Organization in the early 1970s, protracted conflict and tensions with Israel, Arab nationalism in the mid-twentieth century, territorial control of the West Bank until 1988, and the Syrian civil war.[6]

Though Jordan does not have specific laws guiding their management of refugees, there are several agreements and cognate laws that have informed Jordan's evolving refugee policies. In the remainder of this subsection, I will discuss a 1998 memorandum of understanding (MOU) with the UNHCR and Jordan's 1973 Law on Residence and Foreigners' Affairs, as these two formal documents do guide aspects of refugee management in Jordan. However, as I'll highlight in many instances throughout this book, Jordan's refugee management practices have multiple other influences.

An MOU signed between Jordan and the UNHCR in 1998 is a major, guiding document for Jordan's refugee policies (Saliba 2016).[7] This MOU, however, does not pertain to Palestinians, who are under the mandate of the UN Relief and Works Agency for Palestine Refugees (UNRWA; see the following section "Palestinian Displacement in Jordan" for details). The MOU echoes approximately 70 percent of the principles, definitions, and charges of the 1951 convention and the 1967 protocol (Bidinger et al. 2014). As stated in both the MOU and the convention, individual states and the international community are obligated to safeguard asylum seekers and refugees. Article 5 of the MOU declares that Jordan will treat asylum seekers in accordance with international standards, including being "humanitarian and peaceful" and observing the principle of non-refoulement, which, again, is a cornerstone principle of the international refugee regime.[8] The MOU defines a "refugee" in the same way as in article 1 of the 1951 convention, but without geographic or time limitations as in the 1967 protocol.

One substantial difference between the MOU and international standards pertains to the durable solutions. Article 5 of the MOU stipulates that "a durable solution, be it voluntary repatriation to the country of origin or resettlement in a third country," must be found. Article 10 states that "in order to find durable solutions and to facilitate voluntary repatriation or resettlement in a third country," refugees are exempt from overstay fines and departure fees. Local integration is pointedly excluded as an option in the MOU (as it was also in the Syrian 3RP). Thus, since 1998 all refugees in Jordan are formally excluded from integration.[9] This MOU policy is remarkable, not only because it excludes one of the three durable solutions but also because de facto integration is incredibly common in Jordan. In other words, the reality that millions of refugees in Jordan have been informally integrated stands in stark contrast to the stated options in the MOU.

Underwriting some of Jordan's refugee governance is the 1973 Law on Residence and Foreigners' Affairs. This law stipulates regulations for foreign nationals (defined as people not possessing Jordanian nationality) to enter, exit, and acquire residency in the state (International Labour Organization 2015; Sadek 2013; Saliba 2016; Stevens 2013). Article 6 requires that those entering Jordan through unofficial points, such as asylum seekers coming through unofficial border crossings, must present themselves to officials within forty-eight hours of their arrival in order to be registered. Article 31 grants the Ministry of the Interior the power to determine the legality of entry and to deport non-nationals who are in violation of Jordanian laws. The 1973 law applies to all foreigners without distinctions between refugees, asylum seekers, and other migrants. However, this law does grant slightly more open and favorable treatment to Arab non-nationals from neighboring states. For example, the fee for residency permits (articles 23 and 30d) is waived for Arabs from other Arab-majority states. Theoretically, this 1973 law continues to guide regulations pertaining to residency permits and visas for refugees, but subsequent practices have been implemented, and some are in contradiction to the 1973 law. For example, the 1973 law allowed Iraqi nationals visa-free entry into Jordan, but Jordan implemented a new practice in 2005 that required Iraqis to obtain visas prior to their arrival. This new requirement is a restrictive bordering practice that has had the effect of weakening Iraqis' ability to cross into Jordan to seek refuge in the wake of the 2003 US invasion of Iraq and the rise of ISIS/Daesh in 2014.

Palestinian Displacement in Jordan

Palestinian refugees have the unfortunate distinction of being the most protracted refugee population in the world (Takkenberg 2010). The declaration of the state of Israel in 1948 and the war that immediately followed resulted in the forced displacement of 711,000 Palestinian Arabs (out of a total population of 900,000; UN 1950). The 1967 war and the Israeli occupation of the West Bank and the Gaza Strip that year created a second wave of approximately 280,000 Palestinian refugees. As of mid-2019, there are more than five million registered Palestinian refugees residing in Jordan, the West Bank, the Gaza Strip, Lebanon, and Syria, while there are hundreds of thousands of unregistered Palestinian refugees scattered across the globe.

Of the more than five million Palestinian refugees registered with UNRWA, about 40 percent live in Jordan.[10] As of September 2019, 2,206,736 Palestinian refugees are registered with UNRWA in Jordan. There are also many Palestinians in Jordan who are not registered with UNRWA, some because they don't need assistance, others because they lack proper documentation, and others still because they are ineligible for refugee status. While the exact number of Palestinians in Jordan is unknown, it is commonly estimated that around 50 percent of Jordan's total population of just over ten million people are Palestinians. Some estimates, however, figure the Palestinian population as high as 60–65 percent of the total Jordanian population (El-Abed 2014, 86).

The majority of Palestinians in Jordan were born in Jordan and have never been to Palestine. These Palestinians did not experience the *act* of forced displacement but are descendants of family members who were forcibly displaced in 1948 or 1967. Most Palestinians in Jordan live within about 30 miles of Israel/Palestine, but Israel prohibits the majority of them from crossing the border.

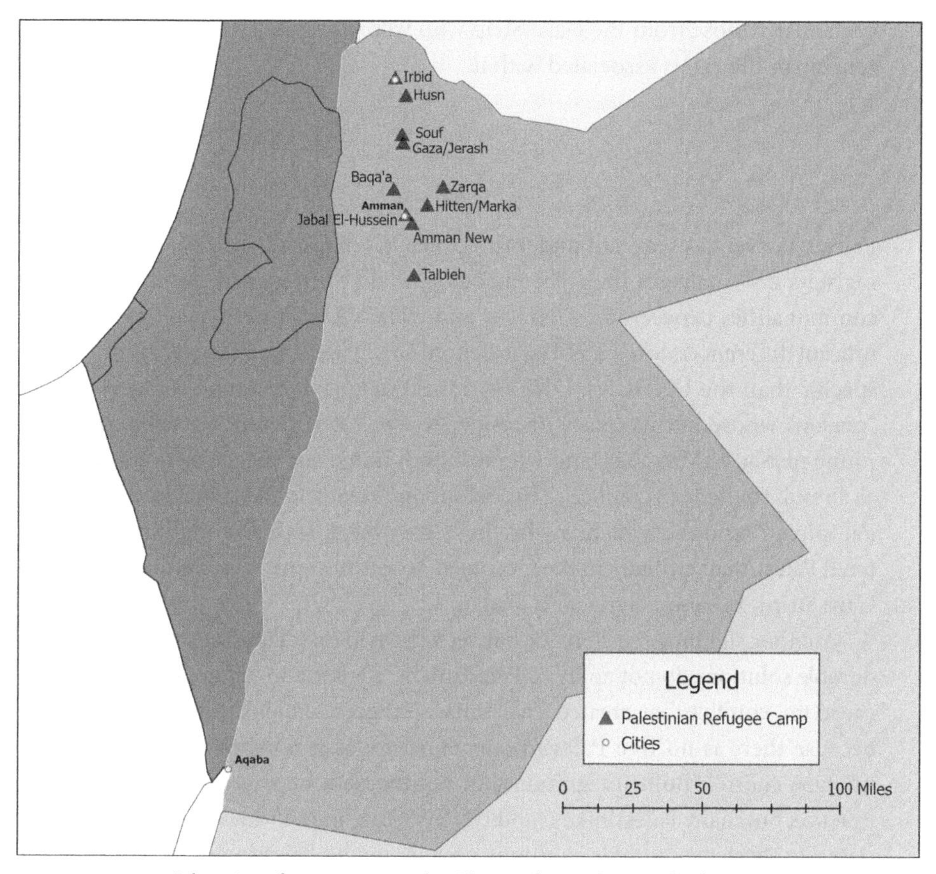

FIGURE 3.2. Palestinian refugee camps in Jordan. This map locates the ten official UNRWA camps.

Palestinian refugees have been living in Jordan for as long as seventy years and constitute at least half of the total population of Jordan. They have been, perhaps unsurprisingly, integral to all aspects of Jordanian society, economy, and politics. A large majority of Palestinian refugees are Jordanian citizens, and 82 percent of registered Palestinian refugees live within towns and cities rather than in camps (figure 3.2). Many Palestinians are highly educated, and they have dominated the private economic sector. While Palestinians are central to Jordanian society, they do face some significant challenges as refugees and have been subject to exclusionary practices and discrimination (Anderson 2005; Massad 2001; Ryan 2011, 40; Milton-Edwards and Hinchcliffe 2009). For example, Palestinians are barred from serving in the highest security and defense positions, and they are underrepresented in the Parliament and Cabinet (Massad 2001, 258).[11] Moreover, poverty is an issue for many Palestinians, particularly those from the Gaza Strip who were not granted Jordanian citizenship or the rights associated with it.

UNRWA AND THE ABSENCE OF THE DURABLE SOLUTIONS

Palestinian refugees are not under the UNHCR's mandate. Instead, UNRWA registers and manages their protections and aid.[12] Though there are some commonalities between the UNHCR and UNRWA mandates, there are significant differences, too.[13] UNRWA's definition of Palestinian refugees is more specific than the UNHCR's. UNRWA (2022) defines Palestinian refugees as "persons whose normal place of residence was Palestine during the period 1 June 1946 to 15 May 1948, and who lost both home and means of livelihood as a result of the 1948 conflict." This definition extends to the children of male Palestinian refugees, which is why there are now at least five million registered Palestinian refugees in the Occupied Territories (the West Bank and the Gaza Strip), Lebanon, Syria, and Jordan.[14]

Another significant difference between the two mandates is that the three durable solutions do not apply to Palestinians. Repatriation is impossible because the conflict that created Palestinian refugees remains unresolved and because there is no free Palestine to return to, as it remains under Israeli military control. Both integration and resettlement have been common in practice, but many Palestinians (and UNRWA) are morally and politically opposed to these two durable solutions because giving Palestinians permanent residency status in a state other than Palestine could undermine their legal "right to return" (*al-awad*). The "right to return" to Palestine is enshrined in the 1948 UN Resolution 194. It states that "refugees wishing to return to their homes and live at peace with their neighbors should be permitted to do so at the earliest practicable date, and that compensation should be paid for the property of those choosing not to return and for loss of or damage to property which, under principles of international law or equity, should be made good by the Governments or authorities responsible." In order to uphold their "right of return," several Arab governments that host Palestinians, including Lebanon, Syria, Libya, and Jordan since the late 1960s, have refused to grant Palestinian refugees permanent status or citizenship in their states. This practice, while symbolically significant in asserting their "right to return" to Palestine, has rendered these Palestinians stateless people without the human and political rights that typically come with citizenship.[15]

The protracted Palestinian refugee situation has shaped reactions and policies toward other refugee populations across SWANA (Chatelard 2010a,

41). Isotalo (2014) argues that a "fear of Palestinization" in Jordan, Lebanon, and Syria (the states that host the most Palestinian refugees) has directly impacted these governments' approaches toward all subsequent waves of mass displacement. Yet, while these states may be reticent to accept new refugees because they "fear" these populations might become more or less permanent refugees like the Palestinians (Francis 2015), Jordan, Lebanon, and Syria have been largely open to allowing in other forcibly displaced people, most notably Iraqis and Syrians.

LABELS AND CATEGORIES OF PALESTINIANS IN JORDAN

Palestinian refugees in Jordan are grouped into two main categories. The larger group is the Palestinians who were forcibly displaced as a result of the 1948 Arab-Israeli war and who are prevented from returning to what is now the State of Israel and the Occupied Palestinian Territories. Of the 711,000 Palestinians who were displaced in 1948, about 70,000 crossed the border eastward into Jordan (Brand 1995, 47). These refugees and their descendants (on the patrilineal side) are often referred to as "West Bank Jordanians." They are Jordanian citizens with the same political and human rights as "East Bank Jordanians" (who are also referred to as "Transjordanians").[16] Being refugees who also have citizenship gives them the rather unique and contradictory status as "refugee-citizens." Typically, when a refugee is granted citizenship, their status as a refugee ends. Moreover, it is common that refugees reject the label of "refugee" because of its negative connotations. But for many Palestinians, to self-identify as "refugees," regardless of their citizenship status, is an important political label that signifies their enduring displacement and steadfast intention to return to Palestine (El-Abed 2014, 90).

The second major category of Palestinians in Jordan is "displaced persons" (*nazeheen*). These are Palestinians who arrived in Jordan as a result of the 1967 Six-Day War. During this brief war, the Israeli military occupied the Jordanian-controlled territory of the West Bank, the Egyptian-controlled Gaza Strip and Sinai Peninsula, and Syria's Golan Heights. Although it is illegal under international law, the Israeli military continues to occupy the West Bank and the Golan Heights and maintains military control over all of the Gaza Strip's borders. As a result of the 1967 war, approximately 80,000 Palestinians from the Gaza Strip sought refuge in Jordan (Shiblak 1996). For some of these Palestinians, Gaza was their original home, while for others it was where they were displaced to in 1948. For this latter group, the move to Jordan was their second displacement in twenty years.

Unlike the 1948 refugees, neither UNRWA nor the Jordanian government labels 1967 displaced Palestinians as "refugees," nor were these Palestinians granted citizenship in Jordan. Based on UNRWA's founding mandate, a Palestinian refugee is someone displaced as a result of the "1948 Arab-Israeli conflict." While it is conceivable that the 1967 war was a continuation of the 1948 war, nevertheless, Palestinians displaced from the Gaza Strip into Jordan in 1967 are officially "displaced persons." Even though they are not labeled "refugees," these Palestinians are under the UNRWA mandate because in June 1967, UN Resolution 2252 expanded UNRWA's initial mandate in order to provide these Palestinians with "humanitarian assistance, as far as practicable, on an emergency basis and as a temporary measure."[17]

The legal status and rights of Palestinians from the Gaza Strip remains a palpable issue in Jordan. There are about 140,000 stateless Gazan Palestinians in Jordan (Chatelard 2010b). These Palestinians, many of who are first-, second-, and third-generation "displaced people," live in limbo as stateless people. They are governed in part as foreign nationals as defined in the 1973 law and thus have limited basic rights and minimal access to government services (El-Abed 2005, 4, 8). The Jordanian government issues them temporary two-year passports and identification cards that label them as "displaced people." Their residency permits are also temporary, and they are restricted from several areas of employment, including public-sector jobs and legal practice (Marshood 2010, 67–68; Tiltnes and Zhang 2013, 32). Palestinians from Gaza are not offered public tuition for higher education, they do not qualify for national health coverage (except children under the age of six), and they do not receive welfare benefits. These Palestinians can own only one car and cannot be the sole owner of a business. As a result of their lack of rights and limited access to governmental services, education, and employment, these Palestinians face poverty at much higher rates than Palestinian Jordanian citizens (Tiltnes and Zhang 2013). However, through UNRWA, these Palestinians can and do receive social services, like health care and primary and secondary education.

Many Palestinians—both those with citizenship and without—express concerns that the treatment of the 1967 refugees is unfair, exclusionary, and harmful. Some Palestinians I interviewed asserted that the government should grant these "displaced" Palestinians citizenship, which would improve their quality of life. Jordan, however, is unlikely to do so for several reasons. First, granting these 140,000 Palestinians citizenship would enfranchise them with voting rights that could shift the political power that the East Bank/Transjordanian citizens currently hold over West Bank/Palestinian citizens. Second, these refugees symbolize and remind the world of Israeli aggression and its continued refusal to abide by international laws. In other words, these stateless people serve as symbols of the "right to return." And therefore, granting the 1967 Palestinians citizenship, which would mean permanent legal status outside of Palestine, could undermine the Palestinian independence movement and their goal to return.

The different labels and categories of Palestinians in Jordan—as refugees in 1948 and displaced persons in 1967—represent a substantial shift within Jordan's refugee policies and practices. They also highlight a clear example of how the state government includes some people and excludes others. Crucially, that shift in practice and its exclusionary effects has had significant impacts on Gazan Palestinians, rendering them stateless and creating countless obstacles in their everyday lives.

SHIFTING REGIONAL AND NATIONAL INTERESTS

Over the past seventy years, since the mass, forced displacement of Palestinians in 1948, Jordan's treatment of and approach toward Palestinian refugees has been intertwined with its cross-border regional geopolitics and its domestic nation- and state-building projects.

The Arab nationalist movement, as I discuss in detail in chapter 4, has had many forms, but at its broadest, it is a regional movement that attempted to unite Arabs across the imperial-drawn borders of the fallen Ottoman Empire. From the early 1920s until 1950, Jordan's King Abdullah I sought to create a united Arab territory from the British- and French-mandated states of Jordan, Syria (which included Lebanon), and Palestine (Pappe 1994; Abu-Odeh 1999; Porath 1984).[18] While still under the British and French mandates, Abdullah lobbied British and French leaders to redraw the borders in order to merge Syria, Palestine, and Transjordan into a single state of "Greater Syria," known as *Bilad al-Sham* in Arabic. In 1933, after the death of the king of Iraq (who was Abdullah's brother Faisal), Abdullah sought to incorporate Iraq as well into a newly configured territory of Greater Syria.

Abdullah's attempt to unite these territories was as much about anti-imperialism and uniting the Arabic-speaking majority as it was about economics. Jordan contained no known valuable natural resources, major cities, or universities. However, the Syrian mandate contained the cities of Damascus, Aleppo, Homs, and Beirut; Palestine included Jerusalem (*al-Quds* in Arabic); and Iraq had Baghdad. All of these cities were of immense cultural, historic, and economic value. Syrian, Palestinian, and Iraqi territories also had productive agricultural land and access to the Mediterranean Sea and/or the Arabian (Persian) Gulf.

Abdullah I never succeeded in merging and controlling Greater Syria. This was largely because other Arab leaders resisted uniting under his leadership and preferred to begin their own processes of producing their own independent states and nations. However, Abdullah I did expand Jordan's territory westward into one territory of paramount importance.

During the 1948 Arab-Israeli war, Jordan's "Arab Army" conquered the West Bank and, in December 1948, officially annexed it. The Hashemite Kingdom began purporting that Jordan was "the representative" of Palestine and Palestinians. The approximately 720,000 Palestinians living in the West Bank (including long-term residents as well as recently displaced people from other parts of Palestine) were granted Jordanian citizenship. The population of Jordan just prior to the annexation was between 375,000 and 440,000 people. With the annexation of the West Bank, Jordan's population skyrocketed to well over one million people practically overnight (Massad 2001, 233; Brand 1995, 47). Acquiring the West Bank was incredibly beneficial to the burgeoning Jordanian state. Many of the West Bank Palestinians who instantly became Jordanian citizens were highly educated, and many had skills and expertise to help build the Jordanian state and grow its economy. In addition, annexing the West Bank gave Jordan control of important religious and tourist sites in and around Jerusalem, along with other well-established cities, institutions, and valuable agricultural land. While this unification legally included Palestinians in Jordan, it was an unequal process. The Hashemite monarchy retained its position as the undisputed leader, and political and governmental power remained centralized in Amman. Thus, it is not surprising that the Jordanian annexation of the West Bank was not widely celebrated by Palestinians.

Numerous leaders, including the Palestinian representative to the Arab League, the mayors of Jerusalem and Nablus, and the Palestinian Liberation Organization (PLO), openly declared their opposition to Jordan's annexation of the West Bank (Massad 2001, 228). Likewise, many West Bank Palestinians rejected the territorial appropriation, seeing Jordan as an occupying power and as undermining their independence (Abu-Odeh 1999, 57; Gandolfo 2012, 16). Despite this strong and widespread opposition to Jordan's annexation of the West Bank, over time, it was normalized, and most Palestinians came to accept it.

The PLO, formed in 1964, maintained its headquarters in the West Bank until it was exiled in 1967 with Israel's occupation. It then moved eastward and reestablished its headquarters in Amman. Soon, the PLO began to create its own quasi state within Jordan. It had its own small territories (often based in refugee camps), its own schools, and an armed militia. In September 1970, frustrated by the growing strength and independence of the PLO, the

Jordanian army began to deport PLO militia. A civil war, known as Black September, soon erupted between the PLO and the Jordanian army. The war continued into the summer of 1971 and ended with the PLO's defeat. Though estimates vary greatly, as many as 20,000 people were killed during the war, the majority of whom were Palestinian. Areas of Amman and large portions of several Palestinian refugee camps were destroyed (Massad, 2001, 244–45).

This war and a disdain for the PLO propelled the formation of an East Bank/ Transjordanian nationalism, which was pointedly against the inclusion of Palestinians within Jordan. Instead, this national identity sought "Jordan for Jordanians" (Massad 2001, 247; Abu-Odeh 1999, 22).[19] While Transjordanian nationalism had existed to some degree prior to the 1970–71 Black September war, the Transjordanian movement grew in size and influence as a result of the conflict (Farah 2008, 86). Soon afterward, during the 1974 Arab League meeting in Rabat, Jordan rescinded its decades-long self-proclaimed status as "the representative" of Palestine and Palestinians. At this meeting, the league's members, including Jordan's King Hussein, declared that the PLO was now the "sole legitimate representative of the Palestinian people."

Fourteen years later, in August 1988, King Hussein officially renounced Jordan's claims and responsibilities to the West Bank. His decision to do this was in part due to the Palestinian independence movement that had erupted in the West Bank against Israel in 1987, which would later become known as the First Intifada. In addition, while Jordan maintained *legal* control of the West Bank until 1988, in reality, Jordan had lost all control of the territory with Israel's occupation of it in 1967. In renouncing Jordan's annexation, Hussein formally recognized the territorial separation of Jordan from the West Bank and denationalized approximately 700,000 Palestinian Jordanians living in the West Bank (Abu-Odeh 1999, 271; Tal 1993; Gandolfo 2012, 68). Palestinians who have lived in the Israeli-occupied West Bank have been stateless ever since the renunciation.[20] King Hussein's action represented a major shift away from Jordan's forty-year-old discourse and practice of attempting to unify Jordan and Palestine, and a refocus on state- and nation-building within the borders of the British Mandate of Transjordan. The Jordanian monarchy now believed that "Jordan is Jordan and Palestine is Palestine" (Fruchter-Ronen 2013, 290; Abu-Odeh 1999, 194). Jordanian state institutions slowly published new maps to reflect the change of Jordan's territorial extent. By 2003, the maps sanctioned by the Jordanian government no longer included the West Bank as part of its territory. Instead, Jordan's territorial extent ended at the Jordan River (Culcasi 2016). Yet maps that include Jordan and the West Bank (and sometimes even the entirety of Historic Palestine) as one territory still circulate in older atlases and map collections.

Initiated by the Hashemite monarchy, a period of nation-building began soon after Jordan's disengagement from the West Bank (Susser 1994, 212; Ryan 2010). The "Jordanization" process sought to create cohesion and unity of people within Jordan's territory, particularly between East Bank Jordanians and West Bank Palestinians. One component of the Jordanization process was a public relations campaign, released in 2002, named "Jordan First." As the name clearly indicates, the goal was to prioritize Jordan and a Jordanian national identity, in contrast to the former priority of uniting Jordan and Palestine. The Jordan First campaign had its own logo, a blocky outline drawing of three hands clasped together, with each hand in one of the colors of the Jordan flag: green, red, and black. The different-colored hands are linked so as to form the shape of the Jordanian state-territory (Culcasi 2016; Milton-Edwards and Hinchcliffe 2009, 65–66). Several years later, the monarchy modified the "Jordan First" campaign and relaunched it as "We Are All Jordan." They reused the map-logo image but changed the accompanying text. This map-logo, with both labels, can still be found in different public and private spaces in Jordan today (see figure 7.2).

Syrian Displacement in Jordan

As of September 2022, Syrians are the largest group of forcibly displaced people in the world. There are 6.7 million Syrians internally displaced within Syria and 6.6 million more displaced in other states.[21] The fleeing of Syrians into neighboring states—namely, Jordan, Lebanon, and Turkey—has been exponentially larger in volume than Syrian migration into Europe or the US (Hoffmann 2016b, 192–94). Jordan, Lebanon, and Turkey were quite open to Syrians initially, but each state has implemented restrictive bordering practices and policies to hinder Syrian refugees from crossing into their states' territories.

The Syrian war erupted in March of 2011. As the "Arab Spring" swept through the SWANA region, Syrians in the southern part of the state began to protest the oppressive and often brutal treatment of civilians by the Assad regime and its security forces.[22] President Assad did not yield to the Syrian uprising as Mubarak had in Egypt and Ben Ali had in Tunisia, and soon a full-scale war was destroying Syria and devastating the lives of millions of Syrians. Syrians from many different backgrounds participated and supported the revolt against Assad. As Yassin-Kassab and Al-Shami (2016) detail, the educated and uneducated, secularists and moderate Islamists (58), Arabs, Kurds, Christians, and Muslims (65) came together during the revolt. Only as the war went on did it became increasingly sectarian (108). The war continues while I write this in the spring of 2022, but the intensity of fighting has waned. The first and third "durable solutions" have been implemented to some small degree, as some Syrian refugees have moved back to Syria and some have been permanently resettled in third countries. However, the vast majority of Syrian refugees remain in protracted displacement in Jordan, Lebanon, and Turkey, with no "durable solution" available.

An estimated 1.3 million Syrians have fled to Jordan since the Syrian war began in 2011.[23] About half (664,414 as of 2021) are registered refugees with the UNHCR. Of Jordan's population of a bit more than 10 million, these 1.3 million Syrian refugees comprise a sizable new population. Indeed, the impact of Syrian refugees in Jordan has been immense. There have been some tensions between Jordanian citizens and Syrians, and animosity has been directed toward Syrians on occasion (Achilli 2015, 9). However, the majority of Jordanians, both East Bankers and West Bankers, have been welcoming, supportive, and sympathetic toward Syrian refugees.

The concerns that Jordanians have about Syrian refugees rarely involve terrorism, security, or cultural differences, unlike the prevailing discourse in many Global North states. Instead, Jordanians are typically concerned about higher unemployment rates, rent hikes, cost-of-living increases, and reduced social services (Basset and al-Moony 2016). Their concerns are, in other words, over the long-term economic effects of hosting Syrian refugees, particularly a reduction of limited resources (Dardiry 2017, 713). Jordan's economy was struggling well before the Syrian war (Francis 2015), but the war and the 1.3 million Syrian refugees now living in Jordan have exacerbated preexisting economic issues.

Syrians have been subjected to many ad hoc and confusing policies about their mobility, access to services, and rights to work. The Jordanian government's attempts at controlling, monitoring, and ordering Syrian refugees has been disordered but still quite strategic. The government has maintained flexible policies, which allows officials to change their practices to suit different situations and evolving issues. For example, during the first year of the Syrian war, the Jordanian government had a rather laissez-faire, open-border approach toward Syrians entering and staying in Jordan. Syrians did not need visas to cross into Jordan, nor did they need a residency permit to remain lawfully in the country (Saliba 2016; UNHCR 2013b, 2; Bidinger et al. 2014; Clutterbuck et al. 2021; Chatty 2017).[24] However, by early 2013, Jordan and the UNHCR were overwhelmed by the sheer number of Syrians crossing into Jordan, and their registration of refugees became backlogged. As the number of Syrians seeking refuge in Jordan continued to increase, the government implemented restrictive practices and policies that sought to impede their flow across the border. Most notably, this included limiting the number of Syrian refugees allowed in each day (Chatty 2017, 184; Bidinger et al. 2014, 61). Other border restrictions were later implemented—although they have not been consistently observed—to keep out former Syrian state officials, members of the Syrian military, men of military age, Syrians who lacked particular legal documentation, and Palestinians and Iraqis from Syria. Then, in July 2014, the Jordanian government began blocking Syrians at its northern border in larger numbers (figure 3.3). Entries at border crossings went from 1,800 per day in early 2013 to fewer than 200 in late 2014 (Francis 2015, 22). This shift in policy at the border was implemented alongside other restrictions further from the borders (Achilli 2015), like the formal sponsorship practice known as *kefala*, which I discuss in more detail in the next subsection. And then in June 2016, after an attack near Rukban, in the far northeast of Jordan, killed seven Jordanian soldiers, the border was closed entirely. The attack was the

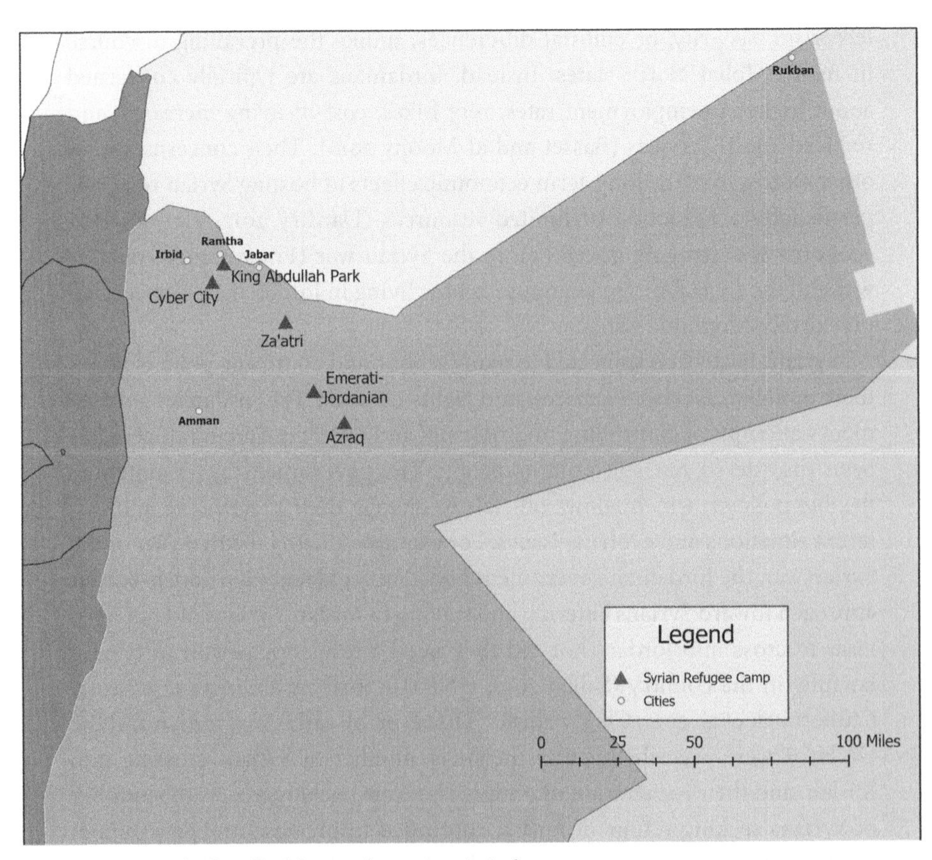

FIGURE 3.3. The five official Syrian refugee camps in Jordan.

official reason for the closure; thus it was purportedly done in the name of national security. However, the broader concerns of how Syrian refugees were impacting the Jordanian economy were an "unspoken" justification (Simpson 2018, 16). Jordan's border restriction and eventual closure (and implementation of *kefala*) mirror the recent practices of several Global North states in prioritizing the protection of its territorial state over the safety of refugees; yet Jordan's restrictive practices did not replace other preexisting, more open practices but instead mixed together.

ENCAMPMENT AND KEFALA

After a bit more than a year of quite unconditionally accepting Syrians fleeing the war, Jordan began the practice of encampment. This was a stark shift in practice, as Jordan had not allowed any camps for Iraqi refugees who had been forcibly displaced after the US invasion of Iraq in 2003. While there are ten UNRWA Palestinian camps within Jordan, those camps are very different form the Syrian ones, as Palestinian camps are open and residents have the freedom of movement (see chapter 7). The Syrian camps, in contrast, function more like detention centers or open-air prisons.

About 6 miles south of the Syrian border is Za'atari, the second-largest refugee camp in the world (after Dadaab in Kenya). The Za'atari camp was built in the desert in July 2012 in response to Syrian forced displacement. The Azraq camp, located in a desolate area about 60 miles east of Amman and 56 miles from the Syrian border, was opened in April 2014 (see figure 3.3). There are three other smaller Syrian camps in Jordan as well: the Emirati-Jordanian camp, Cyber City, and King Abdullah Park. All five of these camps operate like humanitarian centers with basic health, education, and job training services. Yet the camps also function as securitized detention centers by confining and monitoring Syrian refugees within its fences.[25] The encampment of Syrians is happening on a massive scale in Jordan, but only 17 percent (UNHCR 2018a) of Syrian refugees in Jordan live in the camps. The other 83 percent of Syrians are in cities and towns, where living conditions are (for the most part) better and refugees have more independence and dignity.

For Syrian refugees to leave the camps legally and settle in Jordan's towns and cities, they must have proper "bailout" paperwork. This practice was implemented in Jordan about two years into the war and has since been applied in inconsistent ways. In order to obtain a legal "bailout," a Syrian refugee must be sponsored by a Jordanian citizen who is thirty-five or older and married. The Jordanian sponsor is, in theory, responsible for the refugee they are bailing out, including helping them to secure shelter and humanitarian aid.

However, it is not uncommon for Jordanian sponsors to shirk their obligations (Frohlich and Stevens 2015). This sponsorship system is known as *kefala* and is common across much of the Arab world, particularly in the Gulf states. Yet in most other Arab states, it is not applied to refugees but to foreign laborers, and the sponsor is typically a business.[26]

It costs between 360 and 600 Jordanian dinars (US$500–US$1,000) to obtain the bailout paperwork. This cost makes *kefala* prohibitive to many Syrian refugees. For those who have little or no chance of a legal bailout, escaping from the camps illegally becomes a last-resort option. While fleeing the camps to live in towns and cities can offer freedom, opportunities, and dignity, if Syrians leave the camps without legal bailout paperwork, they cannot obtain official identity cards, which means they cannot access UNHCR humanitarian aid (including health and education services) outside the camps. The denial of identification and aid are punitive measures for those refugees who leave the camps illegally, as well as deterrence measures for those who may consider leaving without the proper bailout paperwork. If caught outside the camps and without the proper papers, Syrians may be re-encamped or even deported, both of which violate the principles agreed to in the 1998 MOU with the UNHCR. Regardless of the ramifications of leaving the camps without proper paperwork, Syrians have done so in large numbers (see chapter 7) and thus are living in Jordan's cities and towns without access to UNHCR or state-based humanitarian aid. Syrians I met in Amman and Irbid who did not have proper bailout papers were fully aware of the risks and difficulties they would face outside the camps, yet their lives in the camps had been intolerable and thus fleeing the camps without proper papers was worth the risk for them.

After about four years of punishing former camp residents who left the camps without *kefala*, the government announced a "rectification of status campaign" that issued pardons and amnesty to Syrians who were outside of the camps without proper paperwork. These pardons also applied to Syrians in Jordan who were unregistered and had never been in the camps (UNHCR 2018). Generally, this campaign is seen as a positive one that provided some social and humanitarian support to Syrians who had been residing outside of the camps without the proper paperwork. Yet the "rectification" process, like encampment and *kefala*, is also a way for the government to document, order, and monitor refugees closely, often with the stated goal of national security.[27] The practices of encampment, *kefala*, and rectification each demonstrate that Jordan is confining and monitoring the refugees who are already within its borders. Further, these practices are also a form of strategic deterrence in which Jordan has clearly signaled to *potential* refugees that Jordan is no longer a welcoming place to seek refuge (Nassar and Stel 2019).

BORDER CROSSINGS, CLOSURES, AND REGISTRATION

Syrians have crossed into Jordan at many points along the 235-mile Jordan/Syria border. There are two main crossings on major roads that traverse the border. The first connects the town of Dera'a, Syria, to Ramtha, Jordan; the second connects Nasib, Syria, to Jaber, Jordan. There are as many as twenty-five more remote areas along the border where Syrians have crossed to seek refuge, including in the far northeast of Rukban (Bidinger et al. 2014, 62). Due to "security concerns," the Jordanian government closed the Ramtha crossing in 2011 and the Jaber crossing in 2015, meaning that all subsequent crossings took place at more remote points.[28] In late 2018, both the Ramtha and Jaber crossings reopened, with much celebration and hope of revitalizing trade and human mobility between Syria and Jordan.

After the June 2016 attack near Rukban, as mentioned above, the government closed the entire border and stopped allowing in Syrian asylum seekers at any crossing. This restrictive measure created its own unique territory of displacement, as it trapped 60,000–80,000 Syrians in a zone referred to as "the berm." These Syrians were denied legal entry into Jordan, while also being unable to return to their former homes to the north, and thus, they were stuck in geographic limbo (Magid 2017; Hoffmann 2017; Simpson 2018). They were isolated from aid and services, including the most basic humanitarian aid of medical care, water, food, sanitation, and shelter. According to Amnesty International, as of May 2020, there were still 10,000 Syrians living in the berm.[29] An area just north of Rukban has the largest berm settlement, but there are other smaller ones, like al-Hadalat, with about 1,000 Syrians (Simpson 2018). While the closure of the border clearly hindered the ability of Syrians to seek safety in Jordan, it also affected Syrians already in Jordan. With the border closed, leaving Jordan temporarily to visit family and homes (or what was left of them) back in Syria was nearly impossible, as a safe return crossing back into Jordan would be unlikely. Thus, the border closures kept new Syrian refugees from entering Jordan, while concomitantly trapping other Syrian refugees in Jordan.

Many Syrians have crossed into Jordan undetected, and many of them have chosen not to register with the Jordanian Ministry of Interior (MOI) or the UNHCR. This is why there is a large discrepancy between the assumed number of Syrian refugees in Jordan (1,300,000) and the number who are registered (664,414). Lacking legal registration documentation means that these Syrians are at risk of deportation or encampment. They are also unable to receive humanitarian aid from the UNHCR and the Jordanian state because they do not have the proper paperwork and identity cards. Reasons for Syrians to enter at unofficial points include border closures and the long distances they would need to travel to access the nearest legal crossing. Further, some of the Syrians who are within Jordan's borders have chosen not to register because they may lack legal documentation like a passport. Others have chosen not to register because they may have had ties to the Assad regime, such as having been a solider, and therefore they could be considered enemy combatants by Jordanian officials and put at risk of being deported back to Syria.

Those who have entered legally at one of the two main checkpoints or who have been intercepted by officials after crossing at a remote point are generally brought to a transit center or directly to the Za'atari or Azraq camps. Most transit centers are located just south of the border in Jordan. These centers, like Bustana and Ruwayshid, are run by the UNHCR in conjunction with the MOI and the International Organization of Migration. They provide Syrians with basic needs like first aid, food, and water. At these centers, the UNHCR conducts interviews, adds the refugees to the queue for the durable solutions, and obtains biometric iris scans. From transit centers, refugees are soon transported to a camp, a town, or city, depending on their needs, the type of documentation they have, and whether they have a Jordanian sponsor or not (SNAP 2013).[30]

Those who are brought to a camp are registered with the MOI and UNHCR, if they have not already done so at a transit center. Again, registration renders them eligible for aid in the camps, puts them in the queue for the durable solutions, and documents their legal status as camp residents (and therefore requires them to obtain "bailout" papers if they want to leave legally). Refugees outside the camps must be registered with the MOI to obtain a legal identification certificate and to be registered with the UNHCR in order to receive humanitarian aid and refugee status. To facilitate registration of Syrians not living in the camps (which is about 83 percent of all Syrian refugees in Jordan), numerous registration centers have been established across country, including as far south as Aqaba. A large registration center in Amman was opened in 2013 to meet the high demand and backlog of registrations. Rabaa' al-Sarhan is a major reception and registration center just east of the Jaber crossing and north of Za'atari. It opened in 2014, and in its first six months, more than 49,000 refugees were registered there.[31]

STATUS AND SERVICES IN TOWNS AND CITIES

Outside of the camps, registered Syrian are treated on par with other foreign nationals, including Palestinians who came from the Gaza Strip in 1967.

Syrians are not offered permanent residency, nor do they have legal guarantee of housing, employment, or education.

Syrian children living outside of the camps who want to register for school must have UNHCR and MOI cards. Schools are free, but uniforms, books, and transportation have associated costs, which are prohibitive for some Syrian families. With the influx of Syrian refugees since 2011, the Jordanian public school system became overwhelmed, and educational quality in Jordan is purported to have declined (Ministry of Planning 2017). Many schools have moved to a "double shift" system, with Syrians and Jordanians attending morning and afternoon shifts separately. Some Syrian parents worry immensely that their children are being harassed and discriminated against by Jordanian teachers and their peers, and many are also concerned about the overall quality of their children's education.

Syrian refugees living in towns and cities have access to Jordan's public health care system if they have the required UNHCR and MOI cards. Government subsidies for health care have fluctuated over the past ten years, but as of late 2018, Syrian refugees in Jordan who sought health care in the state facilities paid the same rate as foreigners would, minus 20 percent (UNHCR 2018b, 18). Those Syrians without proper identification documents, however, do not have access to public health care, and thus countless Syrians in Jordan struggle for basic health care, as well as for medicines and treatments for serious ailments.

The Jordanian Constitution (specifically, article 23) states that the *right* to work is reserved for Jordanian citizens. Foreign nationals can work in certain sectors in Jordan if they have residency papers, passports, and documentation from their employers. Syrian refugees, according to article 8 of the 2014 MOU, are allowed to work with a permit. In February 2016, the much-touted Jordan Compact was implemented between the Jordanian government and the EU. This compact allowed for free trade with the EU in exchange for increasing the number of work permits issued to Syrians by the Jordanian government. The compact also required Jordan to open up economic sectors that have long been reserved for Jordanian nationals, like construction and hairdressing.[32] In theory, the compact would curtail the migration of Syrians toward Europe, which was the EU's goal, while growing Jordan's trade economy and allowing Syrians to obtain jobs so that they would become "self-reliant."

Syrians applying for work permits must have valid MOI and UNHCR papers. Although it has been waived at times since 2016, there is a processing fee for the permit (International Labour Organization 2017). And refugees often must wait months to receive the permit. While the Jordan Compact improved employment opportunities for Syrian refugees, the application process is long and difficult, and many organizations and refugees believe it is not working well (Howden, Patchett, and Alfred 2017; Lenner and Turner 2019). Considering the difficulty of obtaining a permit, it is not surprising that many Syrians work in Jordan's towns and cities without legal permits. They do so at the risk of being deported or sent to one of the two main camps, yet the need for paid work often trumps the dangers of illegal employment (Culcasi 2019). Those without a work permit may also be subject to exploitation by their employers, as there is no record or governmental oversight of their employment. Many Syrians I interviewed felt that Jordan's restrictive labor policies are unjust and that they have been denied a basic human right. Syrian men and women, in general, want to work, to provide for their families, and to regain their dignity, but there are few opportunities.

The situation for Syrians in Jordan is protracted and precarious. Eighty-five percent of Syrian refugees in Jordan live below the poverty line (UNHCR 2018). The aid situation in Jordan is dire, as donor fatigue set in years ago, and Jordan does not receive as much international humanitarian aid as they budget to be necessary to care for all the Syrians within their borders. While the immediate and basic needs of *most* Syrians have been *mostly* met,[33] access to educational and job opportunities, as well as health care and the ability to move across borders, are still major problems that stunt Syrians' livelihoods and well-being (Culcasi 2019).

DE FACTO INTEGRATION

As noted above, the Jordanian government avoids referring to Syrian refugees as "refugees," even though the UNHCR has registered hundreds of thousands of Syrians with the legal status as refugees (Molnar 2017, 12; International Labour Organization 2015).[34] Instead, the government refers to them as "guests," and sometimes as "visitors," "irregular guests," "Arab guests," or "Arab brothers." Crucially, these terms (save "Arab brothers") all signify the Jordanian government's position that Syrians are temporary residents who will leave. This terminology reflects the government's "durable solutions" as stated in the 3RP, which exclude the option of permanent integration (see chapter 2).

The durable solutions have failed the Syrian refugees who remain in protracted displacement. Repatriation is dangerous, as Syrians rightly fear retribution from the Assad regime if they return. Syria is also lacking a stable government and economy from which returned Syrians could rebuild their lives.[35] Third-country resettlement in signatory states is possible but quite unlikely. In 2017, 4,989 Syrian refugees left Jordan for third-country resettlement, a marked decline from the 21,000 who left in 2016 (UNHCR 2018a). Further, worldwide, less than 1 percent of refugees are resettled in third countries each year.[36] At that rate, if 1 percent of the 1.3 million Syrians in Jordan were resettled today, the refugee population would shrink by only 13,000 people, leaving 1,287,000 Syrians in protracted displacement.

The Jordanian government has expressly rejected local integration as a durable solution and maintains the discourse that Syrians are temporary guests. However, due to the protractedness of the Syrian refugee situation and the low likelihood of repatriation or third-country resettlement for many Syrians, integration and long-term residency is happening in a de facto form. As Molnar (2017) notes, Jordan has become a place of "permanent temporariness" for Syrian refugees.

In Summary and Looking Ahead

During most of its seventy-six-year history as a postcolonial state, Jordan has been open to migrants and displaced people—Palestinians, Iraqis, Syrians, and others—seeking protection from war and persecution. Jordan has been, in many respects, a "haven" for refugees (Chatelard 2010b), and its practices are commendable, particularly when considering Jordan's lack of wealth, high unemployment rates, and limited natural resources. Jordan is not signatory to the either the UN convention on refugees or the 1967 protocol, but the state is, in many respects, less restrictive than many wealthy Global North states that are party to these international legal instruments.[37]

Since the beginning of the Syrian war, Jordan's rather open and compassionate practices have been mixed with restrictive and repressive measures. In other words, humanitarianism and compassion coexist with repressive restrictions on mobility and human rights. Jordan's refugee practices do not fit into a clear binary of being open *or* closed, nor are its practices either compassionate *or* repressive. Rather, an evolving, unclear, and often disordered system manages millions of refugees in starkly different ways. The ambiguity and malleability of refugee policy and practice is not arbitrary or accidental. Instead, Jordanian policies have been quite strategic and have evolved with broader national politics and regional geopolitics, particularly in relation to Arab nationalism, Israeli occupation of Palestine, and the formation of the Jordanian nation and state.

The state-territory nexus frames many of the evolving politics and policies outlined above. There are countless examples of the reification and reproduction of this nexus in Jordan's refugee policies, like the securitization of its borders with the Syrian civil war and the refusal of citizenship to Palestinian refugees from the Gaza Strip. Yet the state-territory nexus is also sidelined and weakened through other state practices. For example, at a broad level, the ubiquity and normalization of refugees in Jordan challenges the modern state system that frames migration and refugees as aberrant, as exceptions to the norm, and as problems to be fixed through reterritorialization. Refugees are not the exception in Jordan but a formative and quite normal part of Jordanian society, politics, culture, and economy.

4

Pre-imperial and Anti-imperial Territories

Eighty percent of all the refugees in the world are living in the Global South. Global South states are, for the most part, postcolonial states, meaning that they are independent states but that they were created and ruled by imperial powers. A person who is displaced from one postcolonial state into a neighboring one is likely traversing an imperial-imposed border. Indeed, Syrians who crossed their southern border into Jordan and Palestinians who fled eastward into Jordan all crossed borders that were imposed by the British and French after the fall of the Ottoman Empire following WWI. Prior to WWI, while still under the Ottoman Empire (1299–1922), the territories we know today as Syria, Israel/Palestine, and Jordan did not exist but were instead parts of several different Ottoman administrative units that had rather fuzzy and porous borders. After the imposition of the European-drawn territorial states across this region and the implementation of the modern state system, the borders become clearer and harder.

In this chapter, I employ a postcolonial lens to discuss the ways that pre-imperial and anti-imperial territorial practices and imaginings linger in the present, affecting both the Jordanian government's policies toward refugees and refugees' experiences with displacement. These pre-imperial and anti-imperial territorial imaginings, as well as senses of belonging, do not conform to the modern political division of the world but instead are decoupled from states and cross modern borders. Examining the ways that the past lingers in the present is crucial because, as postcolonial scholar Edward Said (1993, 4) reminds us, "there is no way the past can be quarantined from the present."

In the pages that follow, I first discuss the pre-imperial territorial organizing of Southwest Asia under the Ottoman Empire, which was highly structured but maintained porous borders that were greatly open to the migration of diverse peoples. In the second section, I describe the territorial parceling of Ottoman lands in Southwest Asia during and after WWI by the British and French. This late-day imperialism was a complex geopolitical process that drastically altered the Ottoman Empire's territorial divisions and practices and ultimately divided the quite open Ottoman territories into the clearly demarcated territorial states that fit within the modern political order today. Third, I discuss the pre-imperial geographic entity *Bilad al-Sham* (Greater Syria). This is a nonstate-territory that does not exist in today's political ordering of the world, but nevertheless, *Bilad al-Sham* continues to exist in the territorial imaginings of Palestinian and Syrian refugees today and of Jordanian and Syrian leaders in the recent past.

I then focus on the rise of the anti-imperial Arab nationalist movement in the mid-twentieth century, a topic I introduced in chapter 3, and particularly on its construction of the territory of *al-Watan al-Arabi* (the Arab Homeland).[1] *Al-Watan al-Arabi* is an anti-imperial territory that does not fit within the modern, spatial divisions of the world. As such, like *Bilad al-Sham*, it is a nonstate-territory. While Arab nationalism and the idea of *al-Watan al-Arabi* are tenuous today, nevertheless, they remain a part of many Palestinian and Syrian refugees' territorial imaginings and senses of belonging, which have impacted their experiences moving and settling (Shami 1996, 11; Yassin-Kassab and Al-Shami 2016, 8).

Then, in the three sections, I focus on laws and policies of the Jordanian state (and some neighboring Arab states), which have variably emphasized and prioritized Arab interconnections and/or rejected Western laws and norms. In the first of these three sections, I examine major transnational policies and practices pertaining to refugees and migration between Arab states. Next, I discuss the complex reasons that many Arab states are not signatories to the international convention on refugees, which is greatly an act of nonconformity to Western values and ideals. Finally, I develop the discussion on Jordan's cornerstone refugee laws and policies, which I introduced in chapter 3, by discussing the ways in which Arab national discourses and territorial geopolitics have been folded into Jordanian state policies on migrants and displaced people.

Ultimately, in this chapter, I highlight the many ways that pre-imperial and anti-imperial histories, geopolitics, and territories intersect and continue to affect both refugees' senses of belonging and state laws and policies today.

Pre-imperial Territories of Southwest Asia

There is no singular or definitive term that denotes the Ottoman-controlled area of Southwest Asia that encompassed today's states of Syria, Lebanon,

Israel/Palestine, Iraq, and Jordan. Historians of the Ottoman Empire generally refer to this area as the "Arab lands," "Arab provinces," or "Arab territories" of the Ottoman Empire (Hathaway 2008). These terms are all reasonable to use because this region did contain an "Arab majority" or "predominately Arabic-speaking people," but it is important to recognize that there was and continues to be ethnoreligious diversity in the region, including a large Kurdish population (Hathaway 2008, 188; Sluglett 2010, 45–46).[2] Other, more Western-centric terms have also been used to denote this area. For example, in the Middle Ages, the French used the term "the Levant," which translates as "the rising" or "where the sun rises," to refer to this region that, from France (and Europe), is to the east and thus where the sun rises. Terms like "the Fertile Crescent" and "the Middle East" are other common Western, imperial terms that encompass what I am referring to as Southwest Asia. However, neither the Ottomans during their rule nor Arabs in these Arab-majority states today have widely used these Western-centric terms. Arabs often refer to the region of Southwest Asia as *al-Mashriq*, which means "the East" (*al-Maghrib* means "the West" and is also the Arabic name for Morocco). *Bilad al-Sham* is another common Arabic term that, as I will discuss in detail below, translates as "Greater Syria" and refers to a territorial entity dating back to the seventh century that roughly matches the modern states of Syria, Lebanon, Israel/Palestine, Iraq, and Jordan. While there is no perfect term to refer to the former Ottoman lands that are now Syria, Lebanon, Israel/Palestine, Iraq, and Jordan, I have chosen to use the term "Southwest Asia," because it is derived from a continental referent as opposed to being a Euro- or Western-centric term. While the division of the continents has European origins (Lewis and Wigen 1997), a term with a continental referent avoids prioritizing a particular people or power's division and labeling of an area. Further, the term Southwest Asia comprises the "SWA" part of the decolonial term "SWANA" (see chapter 3). Arguably, the Arabinan Peninsula states could also be included in the regional category "Southwest Asia," but because of their different imperial histories, I am not including the Arabian Peninsula states as part of Southwest Asia in this particular discussion. SWANA includes Iran, Turkey, and the whole of North Africa, which is a much greater extent than the majority-Arabic-speaking areas of the former Ottoman Empire.

While many influential empires ruled Southwest Asia and had resounding impacts upon it, it was the Turkish Ottoman Empire who controlled it during WWI and who would lose control of the area to the British and French at the conclusion of the war. The Ottomans, who were Turkish Muslims, ruled Southwest Asia for just over four hundred years, beginning in 1516 when they conquered the area. During this long reign there were many forms and styles of governance, yet Ottoman leaders generally granted local autonomy to their peripheral provinces like those in Southwest Asia. The Ottomans had several different ways of dividing and subdividing their extensive territories into different administrative units, but a system of "vilayets" (provinces) and "sanjaks" (subdivisions within provinces) divided their Southwest Asian territories (Pitcher 1968, 1972). The administrative divisions were systematic and ordered, but these territorial units were not equivalent to the modern, Western state-territorial system. The Ottomans' imagining of territory was not state-centric and did not involve clearly bounded discrete parcels of land. Instead, as Kadercan (2017, 166) explains, the Ottomans saw territory "more as a 'fuzzy' continuum, with no real end-points or beginnings." Elden (2009, 47) echoes this point, referring to Ottoman ideas of territory as greatly "zonal, overlapping, nomadic or fluid."

Under the Ottoman Empire, belonging was greatly tied to social or ethnoreligious communities, rather than to land or territory (Chatty 2010a, 285, 295; Chatty 2017, 186). In other words, senses of belonging were largely deterritorialized, as they were not based on where one was born or lived, or on a notion of citizenship, but on socially defined categories and communities of people (Fábos and Isotalo 2014, 12). Within the extensive empire, there were countless different social and ethnoreligious communities. Indeed, the Ottoman Empire is considered to have been one of the most diverse empires in Europe and Asia (Kieser 2019). In the Ottoman Empire's Southwest Asian territories, people practiced different religions, spoke different languages, and observed countless different customs. These territories had an Arabic-speaking Muslim majority, but nevertheless, there still was quite a lot of linguistic and religious diversity—including Turks, Kurds, Sunnis, Alawites, Greek Orthodox, Armenian Christians, Circassians, Chechens, and Assyrians (Kadercan 2017, 160; Chatty 2017, 188)—from which senses of identity and belonging were greatly derived (Sluglett 2010, 45–46).[3]

During the Tanzimat reforms of 1839–1876, the Ottomans attempted to modernize and Westernize their weakening empire. They implemented uniformity in taxation, centralization of governance, military conscription, land reforms and property rights, and secular law (Kadercan 2017, 29). Central to the reforms was maintaining people's allegiance to the Ottoman Empire as opposed to any particular religious or ethnic community (Kieser 2019, 1). So even as the empire began to centralize, secularize, and implement laws on property in the nineteenth century, the diversity of people and communities remained, but they were now theoretically united under the banner of an allegiance to the Ottomans.

In 1908, the Young Turk Revolution erupted and soon led to major changes within the Ottoman Empire. Of the Young Turks' several goals, one was to elevate the status of Turkish Muslims over other ethnoreligious groups in the empire. In service of this goal, the Young Turks initiated a concerted project of *unmixing* its diverse communities and making Turkey more Turkish, a process that is often referred to as "Turkification." This included several "population exchanges" between the Ottoman Empire, Greece, and Bulgaria (Long 2013, 46; Chatty 2010, 28–29, 38). During WWI, the goal of Turkification was one factor that led to the Armenian genocide, during which approximately one million Armenians were killed. In the largely Arab areas of Southwest Asia during the prewar years, the Young Turks implemented direct Turkish control and suppressed the Arabic language. These new oppressive policies created and fomented Arab grievances toward the Ottomans and helped to fuel a burgeoning sense of Arab nationalism in the Arab-dominant provinces (Sluglett 2010, 45).

The Ottoman Empire's territorial extent shrank between 1878 and 1914, as the Empire weakened. This included the loss of its largely Arabic-speaking North African territories of Egypt, Sudan, Tunisia, and Libya to European powers (figure 4.1). But it was WWI that led to the empire's final demise. During the war, the Ottomans joined the Central powers of the Austro-Hungarian and German empires. After four years of fighting, the Ottoman Empire declared defeat and signed the Armistice of Mudros in October 1918. Yet years before its defeat, as I discuss in the next section below, the Arab-majority Southwest Asian territories of the Ottoman Empire were already within the purview of Great Britain and France and had already been tentatively divided between the two European powers with the Sykes-Picot Agreement of 1916.

In summary, the Ottoman Empire's imaginings of the ordering of territory and people were different from those in the West. The former saw borders as porous and fluid and people as greatly deterritorialized, whereas the latter imagined states to be coupled to territories and people to belong to a singular state-territory. The fact that just over one hundred years ago there was an ordering of territory that greatly differed from the state-centric one that dominates the world today clearly demonstrates just how recently in the scope of history that the Western-centric state-territory nexus has come to divide the globe. The diversity of people and the deterritorialized ideas of belonging that ordered the Arab-majority states under the Ottoman Empire have certainly waned since the empire's division just over a hundred years ago, as the Western imagining of state-territories has been formative in this region since the end of World War I. Yet, as I detail later in this chapter, some aspects of the porous borders and deterritorialized senses of belonging under

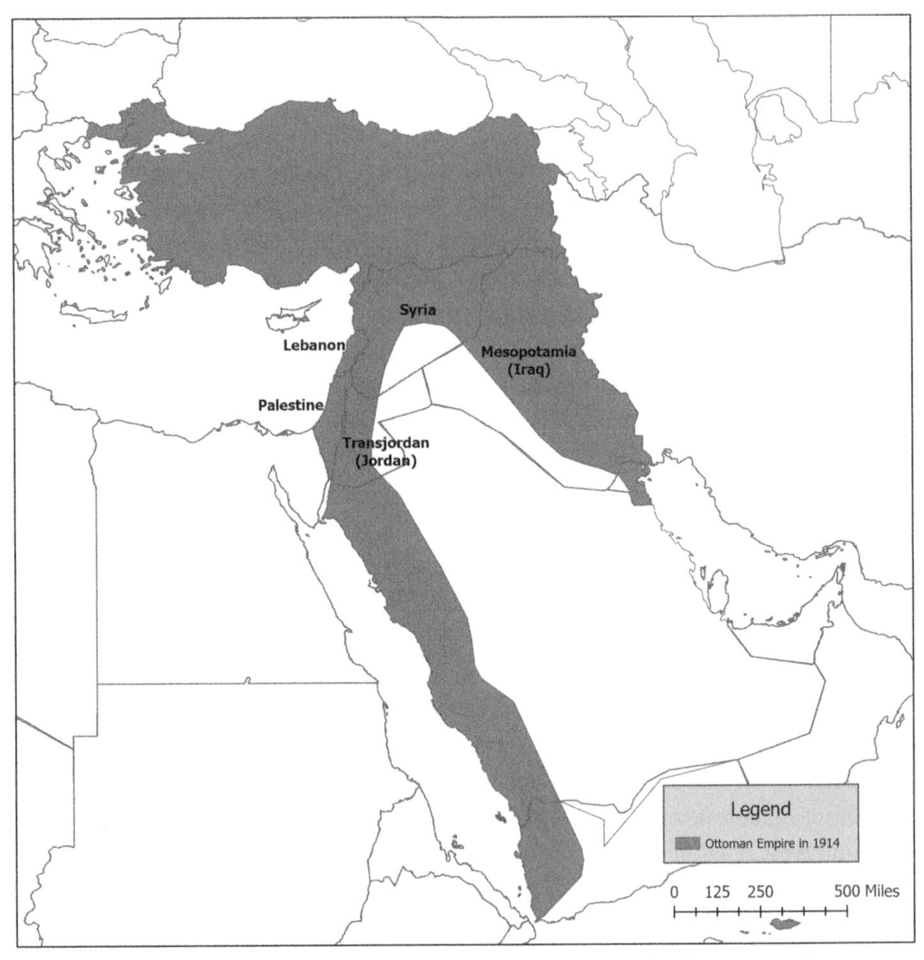

FIGURE 4.1. The Ottoman Empire in 1914 and the post-WWI mandates. The place names on this map refer to the mandates, which went into force in 1923.

the Ottoman Empire have lingered and have continued to shape refugees' imaginings of territory, senses of belonging, and movements (Chatty 2018).

Transition into Territorial, Mandated States

The post-WWI era was a fertile time for the production of state-territories, as major European empires fell and their territories were parceled into different and new territorial entities. During the peace negotiations of WWI, one of the central guiding principles of the Allied victors was that all people and states would have "self-determination," or the right to independence. This principle was central to the American-led vision of creating a "new world order" of independent states that would engage in diplomatic negotiations through the newly formed League of Nations (Smith 2003).

Yet, contrary to this vision, just after the onset of the Paris Peace Conference in January 1919, the Allies adopted article 22 of the Covenant of the League of Nations, which stipulated that a mandate system be enacted in the territories of the former Ottoman, German, and Austro-Hungarian empires. The text of Article 22 reads that in territories "inhabited by peoples not yet able to stand by themselves," an "advanced nation" would take control as a mandatory power. Buttressed by the common Orientalist belief that Arabs and Muslims of Southwest Asia were inferior, violent, primitive, and incapable of self-government (Said 1978, 1993) and fueled by British and French territorial greed, the principle of self-determination was sidelined within the former Ottoman territories. Instead, a new "mandate system" was implemented throughout Southwest Asia, which was a form of imperialism in all but name (Lockman 2004; Adelson 1995; Khalidi 1980).

The geopolitical division of the Southwest Asian area of the Ottoman Empire was a disordered process of reordering territory (Culcasi 2014). In a series of different agreements and treaties from 1916 to 1923, British and French powers created the territories and borders that have evolved into today's states of Syria, Lebanon, Israel/Palestine, Jordan, and Iraq, as well as some borders of Egypt, Saudi Arabia, Turkey, and Iran. Neither the British nor the French had clear plans about how they would rule these new territories, but they did have many geopolitical and economic interests in the region. For the British, protecting their access to India was essential, which meant safeguarding the Suez Canal and access to the Arabian (Persian) Gulf. The British also had concerns about German and Russian power to their east, and therefore they sought to maintain a strong presence in Southwest Asia. For the French, preserving their long-term influence in Christian communities along the eastern Mediterranean was critical. French motivations were partly economic as well. They had invested heavily in building roads, ports, and railways in the eastern Mediterranean and held considerable debt from the Ottoman bank (Tanenbaum 1978; Heffernan 1995). Both the French and the British were also interested in the oil-rich areas in what would become Iraq. For the British, whose Royal Navy had transitioned to oil-fired warships during the war, acquiring control of these undeveloped oil resources was vital.[4]

Many diplomatic negotiations and agreements preceded the final decisions on how to divide Southwest Asia, including the 1916 Sykes-Picot Agreement, the Balfour Declaration, the Hussein-McMahon Agreements, and the 1920 San Remo Conference. The signing and ratification of the Treaty of Lausanne in 1923 marked the official end of hostilities, but it was the 1920 Treaty of Sevres that partitioned the largely Arab areas of the Ottoman Empire into what would become the three British mandates of Iraq, Palestine, and Transjordan, and the French mandate of Syria, which included what is now Lebanon (see figure 4.1). The Treaty of Sevres was signed, but it was never ratified. As Turkey was transitioning into a new state, and Kemal Atatürk rose to prominence as its new leader, he abrogated the Treaty of Sevres and renegotiated it as the Treaty of Lausanne, which Turkey and the European Allies would sign and ratify in 1923. Atatürk's revision of Sevres was not concerned with the Arab-majority areas of the former Ottoman Empire, so the British and French mandates as outlined in the Treaty of Sevres were maintained and enacted. The territories of North Africa from Egypt to the eastern border of Morocco were not part of the postwar negotiations, as the Ottomans had already lost those territories to France, Great Britain, and Italy before the war began. The only Arab-majority area of the former Ottoman Empire not placed under British or French control after WWI was the Hejaz area of the Arabian Peninsula, which contains the holy cities of Medina and Mecca and was under the control of the Hashemite family.

The mandate system implemented a territorial imagining and division that was a radical change from the Ottoman Empire's ideas about territory. The creation of the Iraq, Palestine, Transjordan, and Syria mandates by the British and French laid the foundation for the production of discrete territorial states that fit within the modern ordering of the world. The mandate system of direct imperial control was rather short-lived in these new territories; for the most part it ended after WWII. There were some postindependence modifications to the imperial borders (Saudi Arabia and Jordan made territorial exchanges, the neutral zones between Kuwait and Saudi Arabia have been dissolved, and Israel has occupied Palestinian territory). However, most borders are largely unchanged from the post-WWI negotiations (Burgis 2009, 47, 71). Acceptance of the imperial borders, however, has not been uniform. Indeed, there have been many groups and movements in the mandated states—as in much of the postcolonial world—that have rejected, dismissed, and disregarded Western territorial borders (Burgis 2009, 37, 79). As I discuss below, some of the rejections of the mandate system's borders have been substantial and evolved into movements, while other ones have been more subtle and remained within broader discourses and imaginings of territory and belonging.

The impact and legacies of the WWI imperial borders have been hotly debated (Culcasi, Skop, and Gorman 2019; Abu-Odeh 1999, 271; Joffé 2017; Yassin-Kassab and Al-Shami 2016, 5).[5] Some scholars have argued that this early twentieth-century bout of imperialist state making has affected, if not caused, decades of violence and political rivalry in the region—including the Arab/Israeli conflicts, the Palestine/Israel conflicts, the Lebanese civil war, instability in Iraq, and the rise of ISIS/Daesh. Conversely, others argue that the impact of these imperial borders in creating conflict has been grossly overestimated and that those conflicts are due to internal politics (Schofield 2018; Johnson 2018). Regardless of the role that the imperial borders may or may not have had in leading to conflict, the borders were drawn by imperial powers within their own interests and for their own benefit. The imperial borders and states remain largely intact, and this history has not been forgotten by many people in these new states. Through decades of state- and nation-building, the modern-day territories of Syria, Lebanon, Jordan, Israel/Palestine, and Iraq have become meaningful places of belonging to many people of the Arab-majority areas of the former Ottoman Empire (see chapter 6). However, they coexist with pre-imperial and anti-imperial territorial imaginings that do not mirror the current political ordering of the world.

Pre-imperial *Bilad al-Sham* (Greater Syria)

Bilad al-Sham—translated as "Greater Syria"—is a place name that refers to a pre-imperial territory that includes most of what are today's state-territories of Syria, Lebanon, Israel/Palestine, and Jordan.[6] Some definitions and delineations of *Bilad al-Sham* also include western Iraq, southeast Turkey, Cyprus, and the Sinai Peninsula (Brown 1996, 122; Gelvin 2018, 84). While its boundaries are a bit ambiguous, *Bilad al-Sham* is distinguishable from Mesopotamia to its east, Arabia to its south, and Turkish-dominated areas to its north (to the west is the Mediterranean Sea).

Created in the mid-seventh century as an administrative province during the Rashidun era (632–661), *Bilad al-Sham* has a long and rich history. It became a prominent and powerful area under the Umayyad Empire (661–750) and contained its capital of Damascus. *Bilad al-Sham* ceased to be a formal administrative territory in the late seventh century, and it lost its prominence during the successive Abbasid, Mamluk, Fatimid, and Ottoman empires, which all established their capitals elsewhere. Nevertheless, both its inhabitants and outsiders alike continued to consider *Bilad al-Sham* a coherent and historically important territory, even if ambiguously defined (Philipp 2004, 406).[7] Indeed, 1,400 years after its official disintegration, in the mid-twentieth century, *Bilad al-Sham* re-emerged as a politicized territory that the new Jordanian state and a new political party in Syria both wanted to control.

In the 1930s–40s, King Abdullah I of Jordan, who was ruling under the auspices of the British, pointedly sought to reunite *Bilad al-Sham* as part of

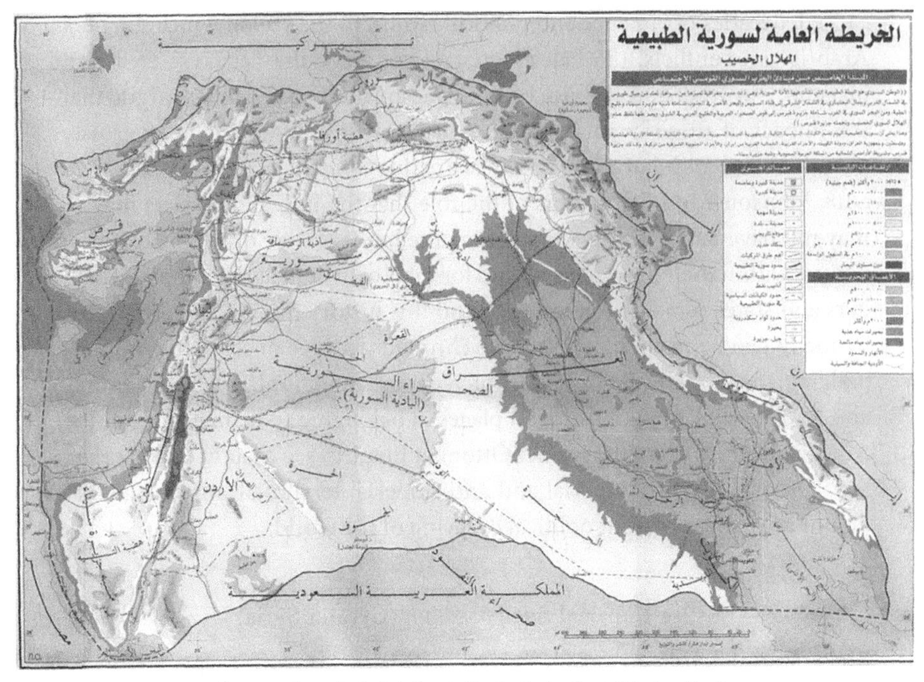

FIGURE 4.2. Map of Greater Syria (*Bilad al-Sham*). By the Syrian Social Nationalist Party.

his wider Arab nationalist goals. As noted in chapter 3, he failed in this attempt, but he did take control of the West Bank in 1948. Likewise, the Syrian Social Nationalist Party, which was formed in 1932, sought to revive the territory of *Bilad al-Sham*. Rejecting the post-WWI imperial borders, the Syrian Social Nationalist Party's territorial imagining of *Bilad al-Sham* was of a natural united territory that included today's Lebanon, Israel/Palestine, the Sinai Peninsula, Iraq, Kuwait, Jordan, Cyprus, portions of northern Saudi Arabia, and southeastern Turkey (figure 4.2). The Assad regime (1971–present) in Syria has also sporadically espoused the idea of recreating *Bilad al-Sham*, asserting that the historical and social commonalities between Jordan, Palestine, Syria, and Lebanon also unite them territorially.

Furthermore, the term *Bilad al-Sham* is still used by many people within the region to refer to a place that has historical and cultural significance to them, even though it does not exist within the modern state system (Chatty 2018, 175, 198). Indeed, on occasion the Syrian and Palestinians refugees whom I interviewed, and who were all displaced from one part of historical *Bilad al-Sham* to another, used this term to refer to a region in which they felt that they belonged. For example, a nineteen-year-old Palestinian woman spoke fondly of the connection that Arabs have with *Bilad al-Sham* and asserted that this historical region, which she defined as consisting of the modern-day states of Egypt, Palestine, Syria, Iraq, and Jordan, was her homeland. Another young Palestinian woman similarly felt that *Bilad al-Sham* was her homeland and continued to explain that she felt connected to all Arabs across the modern-day borders that divided it. A Syrian man in his thirties likewise expressed a strong sense of belonging to *Bilad al-Sham*. He stated, "*Bilad al-Sham* without today's borders is my homeland. Palestine, Jordan, Syria, and Iraq are my countries. They are all my brothers, they are all the same." An elderly Palestinian man talked about the historical geography of Palestine, Jordan, and Syria: "If we go back to the past," he stressed, "it was one country known as *Bilad al-Sham*. It was only later that we were divided into many countries like Palestine and Jordan." Several Palestinians and Syrians likewise noted that it was the British and French who had divided *Bilad al-Sham* after WWI. One Palestinian woman in her forties emphasized that she wants the current borders between Syria, Palestine, and Jordan to be demolished so that *Bilad al-Sham* can be united again. These current territorial imaginings of *Bilad al-Sham* and senses of belonging to it clearly indicate that pre-imperial territorial imaginings continue to exist in the present, even while they do not conform to the modern world order of territorial states.

Anti-imperial *al-Watan al-Arabi* (Arab Homeland)

The origins of Arab nationalism have been intensely debated (e.g., Amin 1978; Antonius 1965; Dawn 1991), but most scholars agree that there were some inklings of Arab nationalism among the elite class in what is today's Syria prior to WWI. However, it was not until the Arab revolt against the Ottomans during WWI that Arab nationalism materialized into a movement. Emir Faisal—a son of Emir Hussein Ibn Ali, the Sharif of Mecca, and the brother to Emir Abdullah, who would become the first king of Jordan—led a newly created Arab army in revolt against the Ottomans. The WWI Arab revolt involved only a tiny portion of Arabs,[8] but it nevertheless sparked the Arab nationalist movement, which would soon strengthen and broaden.

The creation of the British and French mandates after the fall of the Ottoman Empire replaced one dominant power with two others. The Arab nationalist movement therefore pivoted from fighting the Ottomans to revolting against their new European imperialists. Throughout the 1920s–40s, the movement evolved rather slowly. But as each state across SWANA gained its independence, the majority of which did just after WWII, they defined themselves as Arab states and recognized the historical and cultural interconnections across their imperially drawn state borders.[9] Then, in 1948 with the establishment of Israel and the forced displacement of approximately 750,000 Palestinians, the movement proliferated. A sense of solidarity among Arabs across SWANA grew as they witnessed the hardships of Arab Palestinians at the hand of Israel (and with the support of the UK and, later, the US). This concurrence of Arab-majority states gaining their independence from imperial powers, defining themselves as Arab, and witnessing the establishment of Israel as a territorial state in an area dominated by Arab Palestinians, propelled the Arab nationalist movement to new heights (Peteet 1995; Culcasi 2017).

The Arab national movement did lead to the redrawing of a few imperial borders, as a major goal of the movement was to unite Arab-majority territories that had been divided after WWI. Jordan's annexation of the West Bank, as discussed in chapter 3, exemplifies the goal of uniting some of the Arab-majority lands that had been parceled by the British and French. Another example of the attempt to alter the imperial borders and connect Arabs was a brief federation between Jordan and Iraq in 1958. But the most significant territorial materialization of the anti-imperial Arab movement was the creation of a new state called the United Arab Republic (UAR). From 1958 to 1961, Egypt and Syria were united as a singular state, which was internationally recognized and added to the modern state-territorial system with a seat in the United Nations. These three different territorial unifications of Arab-majority lands were all temporary, and the latter two reverted to their earlier divisions in a span of only a few years. Yet each represents the attempt of Arab-majority states to modify imperially imposed borders that divided Arab populations into different state-territories.

In addition to these territorial reconfigurations, the Arab movement of the mid-twentieth century also created and promoted the discursive construction of a large-scale, expansive territory referred to as *al-Watan al-Arabi*, which translates as "the Arab Homeland" (Culcasi 2011).[10] A huge swath of territory stretching from the southern border of modern-day Turkey southward through the Arabian Peninsula and westward across North Africa, *al-Watan al-Arabi* was never a united political entity, but it is a commonplace territorial imagining that symbolizes Arab interconnections across space. Several Arab states and organizations funded the production of maps and atlases of *al-Watan al-Arabi*, and such maps continue to be produced and used across many Arab-majority states. Maps of *al-Watan al-Arabi* often include internal state borders, but there are also many that do not. For example, on a world political map in a Libyan school atlas (figure 4.3) *al-Watan al-Arabi* has no international borders separating its states. Instead, it is represented as one territorial state, akin to the representation of every other independent

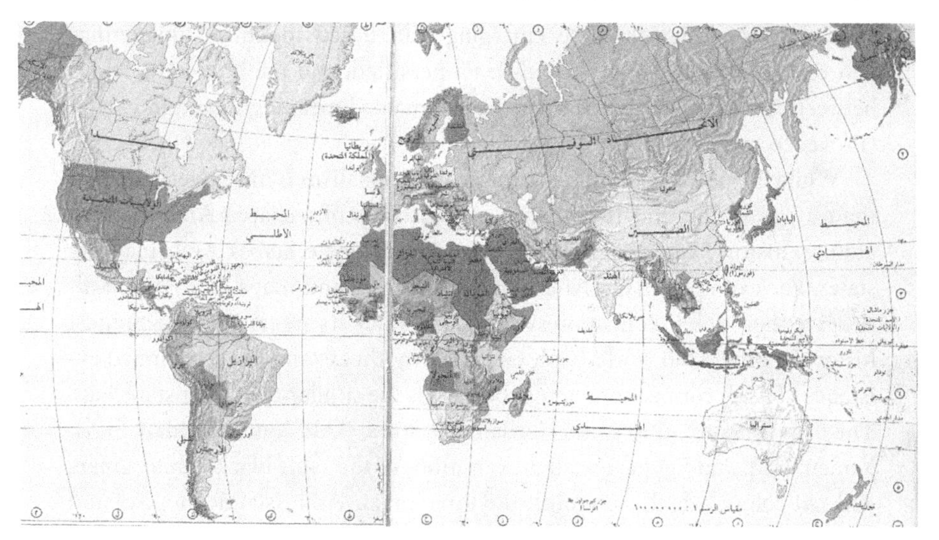

FIGURE 4.3. Map showing the Arab Homeland (*al-Watan al-Arabi*). Excerpt of a world map from the *Educational Atlas for Basic Levels*, published in 1985 by Esselte Map Service Stockholm for the Secretariat of Education, Libya. This political world map, published for Libyan students, shows the Arab world without internal borders. While borders are not shown, each Arab state's name is included.

state across the globe. Regardless of whether or not borders are included on maps of *al-Watan al-Arabi*, such maps represent a territorial imagining that differs from Western imaginings of the division of the world into discrete state-territories and/or world regions.

The Arab nationalist movement was strong through the mid-1960s, but the 1967 Six-Day War dramatically weakened it. During this swift war, Israel invaded and occupied the Gaza Strip and the Sinai Peninsula (both of which had been under Egyptian control), the Golan Heights (under Syrian control), and the West Bank, including Jerusalem (under Jordanian control). The Egyptian, Syrian, Jordanian, and Iraqi forces that fought back were crushed by the Israeli Defense Forces. This military defeat not only led to a loss of significant territory but was also a global embarrassment that seemed to signal that Arab states were weak and incompetent. The Arab nationalist movement, as a result, also suffered a devasting loss. Gamal Abdul Nasser, Egypt's charismatic president and leader of the Arab nationalist movement, died three years after the 1967 war, and with his death the Arab nationalist movement further weakened. The fading of the movement created a political and ideological vacuum across the Arab-majority states; and out of this vacuum individual state nationalisms began to grow strong (Shami 1996, 12). By the 1970s, state- and nation-based identities—like Jordanian, Syrian, and Palestinian—grew stronger. Yet alongside these emerging state-based identities, imaginings and cartographies of *al-Watan al-Arabi* persisted, and the interconnections between Arabs across state borders have remained meaningful and impactful (Provence 2017).

While the power and influence of Arab nationalism is much weaker today than it was in the mid-twentieth century, senses of a shared Arab identity, culture, history, and geography continue to resonate across Arab-majority states. For example, many Arab states use atlases and maps of *al-Watan al-Arabi* in their public education systems, and students are taught geographical histories of the Arab world (Culcasi 2011). Another example of continued existence of Arab connections across borders is the similarity of their state flags. The flags of many Arab states (Jordan, Palestine, UAE, Kuwait, Sudan, Syria, Yemen, Iraq, and Egypt) use slight variations of the same black, white, green, and red colors, which symbolize the three major Arab Islamic empires and the Hashemite leadership of the Arab Revolt of 1916. The Jordanian government flies its own Jordanian flag in countless public spaces, but it also flies a generic Arab flag (figure 4.4) in its southern city of Aqaba. At 131 meters, this is the world's sixth-tallest flag and is starkly visible from the adjacent city of Eilat, Israel. And as I discuss in the next two sections of this chapter, Palestinian and Syrian refugees in Jordan commonly, but not always, feel a sense of

FIGURE 4.4. The Arab flag in Aqaba, Jordan. This is one of the tallest flags in the world and is easily seen from Eilat, Israel.

connection and belonging to some form of *al-Watan al-Arabi* and particularly to fellow Arabs across it. Such cross-border Arab interconnections to an Arab territory and people are not only commonplace but have also affected refugees' migrations, settling, and senses of belonging.

PALESTINIAN REFUGEES' CONNECTIONS TO *AL-WATAN AL-ARABI*

All the Palestinian Jordanians I interviewed, save one, defined themselves as Arab. And nearly all of them said that they felt some sense of belonging and/or historical, cultural, economic, and political interconnections across *al-Watan al-Arabi* and to other Arabs who live throughout it. Some of their feelings of connection were quite strong, others were weak, and many were somewhere in between.

A few Palestinians were clear that they wanted the individual state borders that separate Arab states today to be eliminated, and that in its place a large, united Arab state should be created. A Palestinian woman in her seventies expressed quite warmly that all the neighboring Arab countries felt like her homeland and that the state borders that separate them officially did not divide her homeland in her mind. Another Palestinian woman in her late sixties recognized the borders dividing Arab states but still saw many interconnections across the borders. She wished that "the Arab world would have open borders, just like the EU." When I showed Palestinians maps of the Arab world that contained modern-day borders, many of them used their fingers or a pen to redraw the borders into their imaginings of *al-Watan al-Arabi*. Often, their redrawing created a large, open swath of territory from southern Turkey southward, to include the Arabian Peninsula, and then westward across North Africa.

Belonging to a broad Arab territory was often imagined through connections to Arabs in different states. A Palestinian woman in her twenties told me that even though the Arab world was divided, Arabs themselves are nonetheless "one nation, with one religion, one language, and one soul."[11] Familial terms are quite common in Palestinians' expressions of their imaginings of connectedness and belonging across *al-Watan al-Arabi*. This same young woman, for example, elaborated that Arabs across modern states were all her "brothers and sisters."

Three young men whom I interviewed together expressed their territorial imaginings of unity and senses of interconnections with other Arabs in slightly different ways. One of the young men asserted, "All Arab countries are my countries and all Arabs are my brothers." He continued that if he were to ever live in Algeria, for example, he would feel like he was in his own country because Algerians are also Arabs. One of his friends described the relationship between Arab states as similar to the United States' federal system. He said, "We are combined together like the United States. Each state has its own governor . . . but we are still united as one." The third friend saw such strong connections across borders that he thought government-issued identification cards "should say 'Arab'" on them instead of recognizing individual states. While these three friends expressed slightly different views about their connection to the Arab world, after I showed them the world map (figure 4.3) with *al-Watan al-Arabi* containing no internal state borders, which I often brought with me to interviews, they all asserted empathetically that the borderless map represented a united Arab territory where they felt they belonged.

Territorial imaginings of interconnections across the Arab states are partly produced and framed by migration and travel. Many Palestinians in Jordan have lived in other Arab states and/or have family and/or friends who are living in other Arab states (and elsewhere). The oil-wealthy Gulf states of Kuwait, Saudi Arabia, and the UAE have employed Palestinians of many different skill levels in the oil industry and other economic sectors. Some Palestinians I interviewed had lived in Gulf states, as well as in Libya, Iraq, and Syria. For most of them, having lived in another Arab state strengthened their sense of connections with other Arabs. One Palestinian woman reflected on her years in Libya endearingly, remembering how during Ramadan the different families in her neighborhood would cook dinners for one another. She referred to her Libyan neighbors as her Arab brothers and sisters.

Many Palestinians shared with me their disdain for the imperial borders that divide the Arab world today. A woman in her early twenties told me, "We are one nation. Our origin is Arabia, and we didn't have borders before." A sixty-year-old Palestinian woman spoke of how the Arab world had been one entity in the past, and she opined that the British had divided it. Several Palestinians categorized the British and French dismantling of Arab lands after WWI as an injustice. A Palestinian man in his early forties told me a story of how the British took his family's land near Hebron (in the West Bank) and then built a military airport where their home had once stood. He later talked about how the British had colonized Palestine, but that it was the US, through its support of Israel, who were now responsible for the continued displacement of Palestinians and the injustice of the division of Arab territories.

As noted above, debates on the relevance or legacies of post-WWI imperial divisions, and particularly the Sykes-Picot Agreement, are alive today. There are many economic and geopolitical factors that have fueled tensions and conflict in Southwest Asia since WWI, but it would be misleading to assume that the Sykes-Picot Agreement's effects have not lingered. As my Palestinian interviewees often noted, the Sykes-Picot Agreement and its influence on dividing the Arab lands of the Ottoman Empire is a continuing injustice. A Palestinian man who was the leader (*mukhtar*) of a Palestinian refugee camp in Jordan told me that it was the Sykes-Picot Agreement that divided Arabs and that these divisions were unjust and had caused conflict, particularly in Palestine. A Palestinian woman in her twenties mentioned several times that there once was an Arab nation with no borders, but it had been the Sykes-Picot Agreement and the British mandates that had divided it. She added that Palestinians could not easily move or travel across these modern-day borders. While frustrated by the borders and the immobility they create, she still felt that "Arabs are one nation." Her friend, whom I spoke with at the same time, likewise imagined the Arab world as united and bemoaned the "European treaties" that divided Arabs and the Arab world.

While many Palestinians stressed that it was the imperial borders that divided them, they also commonly recognized that it was the policies and practices of Arab states and leaders that kept them divided. A large majority of the Palestinians I interviewed said that Arab leaders had failed to create political or territorial unity across the Arab world. Some Palestinians expressed hostility toward particular Arab governments for betraying Palestinians. This frustration and sense of betrayal were often directed at Lebanon, Kuwait, and Saudi Arabia, where Palestinians have few rights and restricted opportunities. Several Palestinians referred to Arab leaders as "oblivious" to the needs of Palestinians and as having discriminated against them. A forty-year-old Palestinian man looked at the map of Arab states with internal borders and asserted that all the borders should be removed because the Arab world had once been united; but moments later he told me that Arabs are not united, and he blamed Arab leaders for fomenting tensions and divisions.

Many Palestinians felt quite connected to other Arabs and imagined a united Arab territory; however, some Palestinians felt quite the opposite. A Palestinian man in his early sixties who was employed as a truck driver and had traveled across the Arab world ardently believed that the Arab world was not united. He explained that he felt like a stranger in other Arab states, particularly Saudi Arabia, and that he had been discriminated against by Saudis. One man in his sixties said that he was Jordanian-Palestinian and did not identify as Arab (though he spoke Arabic), nor did he have any connection to other Arabs. A forty-year-old man felt that the Arab world was neither a homeland nor a place where he had any connection. He knew he *should* feel a connection, but he did not. Likewise, the man in his twenties who said "All

Arab countries are my countries and all Arabs are my brothers" also noted differences across the Arab world. He felt that Arab countries were becoming more and more divided. Senses of disunity and division were exacerbated by internal conflicts—like the revolts and wars since 2011. One young man stated, "The recent events in Tunisia, Egypt, Syria, Lebanon, are changing our feelings toward the Arab world. It is divided from within. Even the people are now divided. . . . We are feeling like this is not the same Arab world which was united before the agreements of Sykes-Picot."

Even those Palestinians who felt strongly that the Arab world was divided also frequently expressed some sense of inclusion and belonging as well. In other words, it was common that Palestinians concomitantly imagined interconnections across the Arab world alongside ruptures and disunity. For example, one young man told me that Palestinians were discriminated against in the Arab world and therefore he did not feel united with other Arabs. However, later during our interview he said that "although divided, Arabs are still one nation." The woman in her twenties who felt that Arabs were "her brothers and sisters" later said that she felt connected to Arabs in only some states. She felt that the UAE and Saudi Arabia were not part of her imagined *al-Watan al-Arabi*, but she was more connected to the rest of the Arab world.

SYRIAN REFUGEES' CONNECTIONS TO *AL-WATAN AL-ARABI*

The Syrian refugees I interviewed in Jordan all identified as Arab, with the exception of one woman who was Circassian.[12] For the most part, Arab Syrians told me that they felt a sense of connection to Arabs in other states, as well as feeling that they belonged to a large Arab world. However, as with Palestinians, the depth of their cross-border connections and senses of belonging varied substantially.

Some Syrian refugees in Jordan felt a strong relation with Arabs across modern-day borders. A Syrian man in his thirties expressed his sense of connection in familial terms, stating, "They are all my brethren." A thirty-six-year-old woman told me: "We [Arabs] are one nation and I will never give that up. . . . Even if we have different citizenship, we all love each other." Some Syrians noted that Arabs in different states had different traditions and practices but that they were still connected and united as Arabs. For example, two female friends, aged nineteen and thirty-six, told me that even though Iraqis, Egyptians, and Jordanians had some different traditions, they were all Arabs nevertheless. Likewise, a woman in her mid-twenties said that she did not feel like a refugee in Jordan, because, as Arabs, they were all the same.

Syrian refugees often explained to me that their senses of belonging to a broad Arab world and of interconnections with other Arabs was their reason for coming to and staying in Jordan. A Syrian woman in her late thirties explained that her family had come to Jordan because, in general, "Arab people who are forced to move would usually look for another Arab country to settle in." A young woman in her twenties who was living in the Za'atari refugee camp told me that if she could not return to Syria, she would only live in an Arab country because they shared the same language and because she felt that Arabs were understanding of one another.[13]

A few Syrians felt that the war and their displacement strengthened their sense of Arab unity and connections across borders. Some considered the aid that they received from Arab states and organizations as entwined with Arab values and culture. They were deeply grateful for the assistance they received from other Arabs and Arab states and often commented that this aid cultivated a greater sense of belonging to a cross-border Arab world. For some Syrian refugees, being displaced strengthened ties among Arabs from different states. One Syrian woman, aged twenty-seven, talked about how being displaced had brought Syrians into direct contact with other Arabs and thus created new relations and strengthened abstract ideas of Arab unity. She said, "Before the war, all we knew about other Arab countries was what we saw on TV, but now we are living among them. We are living the reality." Similarly, a woman in her mid-thirties felt that it was the daily act of living together that mattered. She asserted, "We are living among them and at the end all of us are Arab."

Some Syrians expressed great hostility toward Arab governments for their failure in responding to the war and to suffering Syrians. Indeed, many Syrians expressed strong views that Arab governments had "abandoned" them and that this had weakened their senses of belonging to a cross-border Arab world. A fifty-year-old woman was angered that Arab governments "are not helping, they are just watching and counting the numbers of martyrs—today 100 were killed, tomorrow 150, and so on." She felt that if all of the Arab states had intervened, the war would have long been over. A thirty-one-year-old woman asserted that "disappointed" was an inadequate term to describe her feelings. She lamented, "After all the people who were killed, and all the youth who were lost, women who were raped, and children who were . . . [crying], did they [Arab states] not see all of this? It has truly shaken me." Another woman in her twenties was horrified that "Arab countries sat back and watched as Syrians were slaughtered." A thirty-year-old woman was grateful that Jordan and Lebanon had helped Syrians who fled, but she lamented that the rest of the Arab world "abandoned us and drank Pepsi." Syrians often asserted that the Gulf countries (specifically, Saudi Arabia, UAE, and Kuwait) had turned their backs on Syrians by not allowing Syrians in as refugees, by not providing more financial aid, by not pressuring Assad to end the violence, and by exploiting vulnerable Syrian women and girls by marrying them (often temporarily). A woman in her mid-fifties felt as though the Syrian war and Arab states' neglect of Syrians marked the worst moment in all of Arab history.

Many Syrians told me that they had felt a sense of belonging with other Arabs and other Arab countries before the war, but the war and the inadequate reaction from Arab countries had dramatically changed their feelings of cross-border connections and belonging. This war, for many, marked a crucial moment in Arab history when Arab states and Arabs alike stopped looking after one another. A forty-two-year-old woman said, "I love them, I love all Arabs, but I honestly feel that Arab states have let us down." A thirty-six-year-old woman asserted, "I used to be happy that I am an Arab. . . . Now I don't feel the same. . . . The idea of being Arab no longer exists." A Syrian male in his fifties recognized that his thinking about the Arab world had changed. Before the war, he had only ever wanted to live in an Arab state where he would be included, but since the war, he had been ready to leave. One of his friends, in his sixties, elaborated that a sense of Arab brotherhood no longer existed. Similarly, a woman in her twenties said she had once felt connected to Arabs, but she had experienced discrimination in Jordan by fellow Arabs, and this had weakened her sense of connection. A few Syrians mentioned that the closure of the Jordanian border to Syrians fleeing the war had fractured their sense of belonging to Arab states and facilitated their feelings of hostility toward Arab governments, including Jordan's.

Syrian refugees, unlike Palestinians, did not speak of imperial borders, the Sykes-Picot divisions, or European imperialism as important context for understanding their cross-border Arab connections or lack thereof. This is notable but not surprising, as Syrian displacement stems from a recent war within a predominately Arab state, which is a war that does not have a direct relationship to the post-WWI imperialist territorial divisions. The realities of war and displacement have deeply affected Syrians' senses of belonging to a united, cross-border Arab world and their connections to other Arabs. For some Syrians, the war has strengthened interconnections to Arabs and the Arab world, but for many more the war has severed ties. The woman who explained that the daily act of living together with other Arabs in Jordan had fomented connections also asserted that there would have been no war in the first place if there had been Arab unity. For her, and several other Syrians I interviewed, the Syrian war itself, and particularly Assad's slaughter of fellow Syrians, demonstrates quite starkly that Arabs are not united.

Transnational Policies among Arab States

Most Arab states are not signatories to the international convention on refugees or its protocol (a topic I discuss in the next section). There are no legally binding regional agreements on international migration between Arab states, like the rights- and mobility-based agreements that I highlighted at the end of chapter 2. Nevertheless, there are a couple of protocols and declarations in the Arab world that have provided some guidelines and influenced state practices for cross-border movements and forced displacement.[14]

There are two official documents across the Arab-majority states that address refugees specifically: (1) the 1992 Declaration on the Protection of Refugees and Displaced Persons in the Arab World[15] and (2) the 1994 Arab Convention on Regulating Status of Refugees in the Arab Countries. Both have been signed by the members of the League of Arab States, but neither has been ratified, and thus they function as guidelines instead of laws (Sadek 2013). These documents recognize the 1951 convention as well as other international legal instruments on human rights,[16] but they also diverge from the Western-based international laws by asserting the primacy of "Islamic Arab traditions and values." For example, point 2 of the preamble of the 1992 declaration states: "Recalling the humanitarian principles deeply rooted in Islamic Arab traditions and values and the principles and rules of Moslem [*sic*] law (Islamic Shari'a), particularly the principles of social solidarity and asylum, which are reflected in the universally recognized principles of international humanitarian law."[17] Article 5 states: "In situations which may not be covered by the 1951 Convention, the 1967 Protocol, or any other relevant instruments in force or United Nations General Assembly resolutions, refugees, asylum seekers and displaced persons shall nevertheless be protected by: The humanitarian principles of asylum in Islamic law and Arab values." Likewise, the preamble of the 1994 Arab Convention on Regulating Status of Refugees in the Arab Countries reads: "Invoking their religious beliefs and principles deeply rooted in the Arab and Islamic history, which make man such a great value and a noble target that various systems and legislation cooperate to ensure his happiness, freedom and rights." Asserting Arab and Islamic principles in these documents implies that such principles are different from the Western values that frame the international refugee law. This distinction, as I will discuss below, is one reason that many Arab states remain nonsignatories of the 1951 convention and the 1967 protocol.

There are several regional charters and declarations among Arab states that focus on Palestinian refugees specifically, with the 1965 Casablanca Protocol being the most foundational. Adopted by the League of Arab States at the height of the Arab nationalist movement, this protocol established a regional system for the protection of Palestinians, recognizing that Palestinians are Arab brethren who have the right of movement, residency, and work across member states. This international instrument is not framed within Islamic principles but pointedly on Arab ones. The Casablanca Protocol declares that Arab states should treat Palestinians like national citizens, including granting employment and residency rights, issuing travel documents, and granting the right to leave and return like any other citizen.[18] The stated rights of work, residence, and mobility in the Casablanca Protocol are in many ways more progressive and generous than the stated rights of refugees in the 1951 convention (Bidinger et al. 2014, 23).

One way in which the 1965 protocol, the 1992 declaration, and the 1994 convention have impacted migration and refugee practices across Arab states to some degree is through rather loose visa regulations. Indeed, some Arab states have less rigid visa requirements for Arabs from other Arab states, as compared to other non-Arab foreign nationals. This relative openness is duly illustrated in the case of Iraqi refugees' movements after the US-led invasion of Iraq in 2003. Due to their shared "Arabness" and being considered "brethren," Iraqis were allowed visa-free travel into Jordan until 2005 (Gabiam and Fiddian-Qasmiyeh 2017, 6–7; Mason 2011) and into Syria until 2011 (Chatty and Mansour 2011, 59). Studies by Berman (2012), Hoffmann (2016b), and Chatty (2017, 194) all highlight that Syria maintained open borders and laissez-faire policies toward Arab migrants prior to 2011. Syria's openness to Iraqis, as Hoffmann notes, was considered "unmodern" by the UNHCR and was rather "unthinkable" in most Western states. Yet Iraqis in Syria enjoyed a better status and quality of life in that "unmodern" system than they did once the "modern" UNHCR system was implemented in the mid-2000s (Hoffmann 2016b). Likewise, Jordan allowed Iraqis to integrate into towns and cities, refusing to build refugee camps that would likely compromise their mobility and dignity. So, while most Arab states are nonsignatories to the 1951 convention and the 1967 protocol, in several instances, they have policies and practices that have been deemed more humane and open than those in signatory states.

Nonsignatory Arab-Majority States

Only five Arab states of SWANA—Egypt, Yemen, Tunisia, Sudan, and Morocco—have signed the 1951 convention or the 1967 protocol. Eleven—Bahrain, Qatar, UAE, Oman, Algeria, Syria, Jordan, Lebanon, Iraq, Kuwait, and Saudi Arabia—have not.[19] There are two particular explanations for the rejection of the convention and protocol in these eleven Arab states.

At the most general level, the 1951 convention (like the IRR) is Western-centric. It emerged within Western social values, cultural mores, and politics, which included the modern state-territorial system and its goal of each person belonging to one territorial state. The Western-created convention was assumed by its authors to be universal, and as of 1967 it was applied globally, but for many newly independent states, the convention seemed to be a new form of imperialism. Thus, to reject this international legal instrument is an anti-imperialist act of nonconformity to Western values and ideals broadly (Sanyal 2018; Chatty 2017).

Second, Arab and Muslim histories, practices, and values pertaining to migration and asylum are not reflected in the convention (Isotalo 2014). Chatty (2017) argues that people in the "Middle East" [*sic*] have a strong sense of *karam*, which is the duty and obligation to treat refugees, strangers, and guests humanely and generously. She writes that "notions of hospitality and generosity are so important in Middle Eastern cultures as to make it nearly impossible for the state to adopt international refugee law which seems to carry with it the pettiest form of bureaucratic indifference to human needs and suffering" (193; see also El-Abed 2014, 85; Dagtas 2017). Chatty continues, explaining that forced migrants are typically welcomed (or tolerated) as guests in these states, and while the term "guests" implies temporary status, they are frequently welcomed for long durations.

In the case of Syrian refugees, Lebanon and Jordan are nonsignatory states and therefore do not have an obligation under international law to protect refugees. Yet both states have granted millions of Syrians refuge because of "the duty to be generous, to provide sanctuary . . . as kinsmen, business partners, or just fellow humans" (Chatty 2018, 207–8). Turkey, which is predominantly Muslim but not an Arab-majority state, is a signatory of the 1951 convention and the 1967 protocol, but it has maintained the geographical restriction in the 1951 law, meaning that displaced Europeans are the only people Turkey is obligated to "protect." However, as Turkey is the world's number-one host to refugees, most of whom come from outside Europe, its signatory status to protect only Europeans displaced in WWII is not indicative of the realities on the ground (Chatty 2018, 243).

Though often heralded as the standard for good refugee practices, signatory status of the convention or protocol does not determine actual practice. Nor is signatory status indicative of the openness of borders or the quality of services and treatment (Coddington 2018a; Sanyal 2018). Countless signatory states shirk their obligation to protect, as discussed in the introductory chapter, while many nonsignatory states—like Jordan—shoulder the responsibility to grant asylum to people fleeing war.

Jordanian-Arab Policies

Jordan's government has demonstrated an openness of borders and a tolerance for the prolonged stay of refugees for nearly all its history. Despite being a nonsignatory state as well as having a struggling economy and high unemployment rates, the Jordanian state and its citizens pride themselves on their "culture of hospitality" (Davis and Taylor 2013). A leader of the Jordan's Refugees, Displaced Persons, and Forced Migration Studies Center at Yarmouk University explained to me that the reason for Jordan's openness is, quite simply, because Jordan has a "culture and set of values that requires Jordanians to protect those that are insecure."

Senses of Arab interconnections and the anti-imperial Arab nationalist movement have affected the Jordanian government's policies and practices toward refugees.[20] While Arab nationalism is no longer a political force in Jordan, a sense of Arab connectedness across imposed borders does linger (El-Abed 2014, 85; Schwedler 2022, 58–60). This is evident in imaginings of Syrian and Palestinian refugees, as discussed earlier in this chapter; in Jordan's domestic refugee practices, as discussed in chapter 3; and in the Jordanian government's migration and refugee policies with neighboring Arab states.

The ideology of Arab nationalism had, as Chatelard asserted, "provided a rationale for allowing almost unconditional emigration from neighboring Arab states" into Jordan (2010b, 7). Jordan has maintained rather open policies toward Arab nationals for the purposes of tourism, work, study, and investment within Jordan. It has also established bilateral agreements with several of its Arab neighbors to allow visa-free travel and longer-term work-related residency permits (Stevens 2013). The 1973 Law on Residence and Foreigners' Affair (see chapter 3) waived the fee for residency permits (articles 23 and 30d) for Arabs from neighboring states, while other non-nationals are required to pay. It was not until 1984 that Syrians, Egyptians, and other Arabs were required to obtain residency permits for long-term stay in Jordan (Chatelard 2010b, 7). Further, the government's immigration discourse typically refers to Arab migrants from Palestine, Syria, and Iraq not as "Palestinian," "Syrian," and "Iraqi" but as "Arab guests" or "Arab brethren."

The rather open, visa-free policies under which Arabs could enter Jordan became increasingly restrictive after four Iraqi nationals bombed three hotels in Amman in 2005. Three years later, Iraqis were required to obtain a visa prior to entry. Then, with the 2011 war in Syria, visas become more difficult to obtain. According to the Jordanian Tourism Board, as of mid-2021, the citizens of the Arab states of Algeria, Bahrain, Kuwait, Lebanon, Libya,

Morocco, Oman, Saudi Arabia, Tunisia, and the UAE are required to obtain a *free* visa upon entry.[21] With my American passport, I can obtain a tourist visa upon arrival for 40 Jordanian dinars (US$55). Those who need prior approval for a visa and must pay a fee for it include Egyptians, Iraqis, Palestinians, and Yemenis. Syrians, too, are now required to have a visa. Longer-stay visas and residency permits are allowed for Arabs who have job contracts, who are university students, or who are financial investors in Jordan. However, these visas are less common than they were two decades ago (Stevens 2013).

This trend toward greater restriction of mobility into Jordan is also demonstrated by a new, massive border-securitization project. In 2008, Jordan's Border Security Program was announced. In conjunction with the US Army, the Pentagon, Raytheon, and the US-based defense contractor Leonardo DRS Corporation, Jordan has built a security and surveillance system along 160 miles of its border with Syria and along another 115 miles of border with Iraq. This system includes fences, patrol pathways, day and night cameras, sensor-fused barriers, ground radar, both fixed and mobile surveillance watchtowers, and a communication system. The estimated total costs are US$390–US$500 million. The project entered its last phase of construction in 2016, which was five years into the Syrian war and just under two years after the rise of ISIS/Daesh.

In a departure from most of its history, there are now many examples of the Jordanian state denying people mobility to cross into Jordan in search of refuge and of the state's implementation of restrictive practices within its borders (Hanafi 2014, 585–86; Fábos and Isotalo 2014, 12). The closure of the border with Syria in 2016 and the construction of highly securitized Syrian refugee camps both represent a significant shift in Jordan's refugee policies and practices (Hoffmann 2017, 103) and a weakening of cross-border Arab connections. Yet there are also countless examples of openness, generosity, and humane treatment by government offices and leaders, as well as by Jordanian civil society groups and citizens. Jordan, like many Global North states, has engaged in the rhetoric of the need to secure its borders against terrorist threats like ISIS/Daesh. The phrase "safety and security" (*al-a'min wa al-a'yman*) has recently become common among the people in Jordan and their leaders in order to justify Jordan's restrictive bordering practices. Yet Jordan's struggling economy is more commonly pointed to as the driver of the restrictive practices (Zaman 2016). For example, in a July 2012 interview with the BBC, Jordanian foreign minister Nasir Judah explained and justified the recent opening of the Za'atari refugee camp. He stated, "We have delayed the opening of refugee camps for so long because there are family relations, there are intermarriages, there is a history, a demographic history between Syria and Jordan where people have just come in and stayed with relatives. . . ." He explained that Jordan had to open the camps in order to balance "our responsibility to our Syrian brethren" with the "strains on our economy" (Doucet 2012).

In Summary

British and French territorial divisions of Southwest Asia after WWI usurped the pre-imperial territorial openness and replaced it with the modern state-territory system. This new system did not "translate easily" to the territories that were once part of the Arab-majority provinces of the Ottoman Empire (Chatty 2010a, 37). Indeed, the anti-imperial Arab nationalist movement that emerged in the mid-twentieth century was in part a challenge to the imperially imposed divisions of territory. Therefore, when Syrians and Palestinians fled to seek refuge in Jordan, they all crossed imperially imposed borders that had existed for less than one hundred years and that had no pre-WWI significance.

Neither *Bilad al-Sham* nor *al-Watan al-Arabi* exists in today's formal ordering of the world, yet connections among Arabs across these pre-imperial and anti-imperial territories have affected Palestinian and Syrian refugees' territorial imaginings, senses of belonging, and decisions about movement and settling. Many Palestinians and Syrians shared the sentiment that although they have been displaced, they have been displaced from one Arab state to another and thus are not displaced from *al-Watan al-Arabi* or *Bilad al-Sham*. Syrians and Palestinians both commonly felt a sense of belonging and inclusion across the current borders and felt that this sense of connection was a reason to stay in Jordan. Yet that sense of belonging and connection was often tenuous. Indeed, sentiments of disunity coexist with those of lingering interconnections. Both Syrians and Palestinians often expressed a sense that they *should* feel united but did not.

Likewise, pre-imperial and anti-imperial discourses and practices affect some refugee policies across the Arab world and in Jordan. Jordan's liberal visa and residency policies for Arabs are small yet noteworthy materializations of cross-border interconnections. Jordan's borders were rather open and governmental practices were quite tolerant of Arab migrants and refugees until the past decade. In reference to the cross-border connections between Syria and Lebanon, Dionigi (2017) uses the term "thin borders" to describe the pre–Syrian war borders that worked to regulate the movement of things and people, but not to stop these movements. Until the Syrian war, this idea of "thin borders" aptly described Jordan's borders with adjacent Arab states.

In great part, neither the Jordanian state's rather open practices nor the imaginings that Syrians and Palestinians expressed of their connections across borders fit squarely into the state-territorial ordering of the world or the international refugee regime's definition of a refugee or its durable solutions. That these differences exist raises important questions about the relationship between the modern state-territory nexus and forced displacement. For example, are international refugee laws and policies, including the durable solutions and the definition of a "refugee" appropriate for people and situations in which migration is occurring from one part of their broader territorial homeland to another? Considering the Western origins of the current international laws on refugees, can we decolonize our thinking and recognize other territorial forms that do not fit the Western mold?

This chapter has focused on nonstate-territorial forms and imaginings that continue to matter for refugees' experiences and state practices. The example of the lingering imaginings of nonstate-territorial forms of *Bilad al-Sham* and *al-Watan al-Arabi* today is a subtle challenge to the universal idea of the political division of the world into discrete territorial states. This challenge, however, has not usurped the dominant, modernist state-territorial imaginings, a point which I examine in the next two chapters.

5
Hybrid Territories

Pre-imperial and anti-imperial territorial imaginings and feelings of interconnections across modern-day borders, which were the focus of the previous chapter, are impactful in both policies and experiences of displacement. However, they have not usurped the importance of state-territories in general, nor of Jordan, Palestine, and Syria specifically. Instead, cross-border interconnections and pre-imperial and anti-imperial imaginings mix and overlap with twentieth- and twenty-first-century state-territorial divisions.

In this chapter, I focus on the myriad ways that Palestinian refugees in Jordan imagine their senses of belonging and interconnections to both Palestine and Jordan, and likewise how Syrian refugees imagine theirs to both Syria and Jordan. Clearly, there are no Palestine/Jordan or Syria/Jordan independent territorial states today; nor are there any substantial movements or political efforts to construct them. Nevertheless, Palestinian and Syrian refugees in Jordan maintain cross-border connections and imagine their senses of belonging in ways that fold together Jordan with the state from which they or their families were displaced. I refer to such territorial imaginings and senses of belonging as being toward a hybrid territory. Hybridity is a concept that helps us to examine how ideas, things, process, and systems blur seemingly fixed or binary categories. In other words, hybridity inherently merges and mixes things and ideas together in ways that don't fit neat systems, categories, or molds (Foucault 1986; Anzaldúa 1999; Bhabha 2015).

By focusing on the concept of hybrid territories, I tease out the complex ways that the state-territories of Palestine and Jordan and of Syria and Jordan can blur and mix together, and the resulting interconnections and flows, for refugees. Such senses of belonging decenter, transgress, and displace conventional ideas of belonging to one territorial state, albeit without rejecting or erasing the dominant ordering of state-territories. Therefore, hybrid territories are not "trapped" in the state-territory nexus, but they are not fully decoupled from it either. Instead, Palestinians' and Syrians' complex imaginings and senses of belonging simultaneously reify and challenge the spatial ordering of Jordan, Palestine, and Syria, as well as the state-territory nexus in general.

There are two main sections in this chapter: one is focused on the interconnections between Palestine and Jordan and the other on those between Syria and Jordan. Within each of these sections, I discuss the different degrees to which Palestinians and Syrians both imagine and feel a sense of belonging to Jordan and to the state from where they were displaced. This includes their feelings of exclusion, as exclusion is an inherent part of understanding inclusion and belonging. There are countless different configurations of how hybrid territories are imagined and how Syrians and Palestinians feel connected to them, yet they all indicate that meaningful cross-border connections coexist alongside the reification of the state-territory nexus.

Palestine and Jordan

Jordan and Palestine have an intertwined twentieth-century geopolitical history (see chapter 3). After the fall of the Ottoman Empire, the once open lands of Southwest Asia were divided and territorialized to fit Western, imperial designs. Prior to WWI, there were no borders, fences, or checkpoints that divided the area that is now Jordan, Israel/Palestine, and neighboring Arab states. When the League of Nations approved the British mandates of Palestine and Transjordan in 1923, two new discrete territorial entities were added to the political map of the world. The border that was drawn between the two mandates followed the narrow Jordan River, which extends in a north–south direction. While both west and east banks of the river were under the control of the British, the border separated Palestine on the west bank as a Jewish homeland, as promised in the Balfour Declaration of 1917, from Transjordan on the east bank. This imperially created border remains today and has become a highly securitized space that separates Israel/Palestine in the west and Jordan in the east.

During the mandate period (1923–1948), the British administered the mandates separately, but movement across the border was unobstructed. Likewise, from 1948 to 1967, when the West Bank was part of Jordan, movement remained open between the West Bank and Jordan. However, with Israel's occupation of the West Bank in 1967, the border was securitized, and Palestinians living in Jordan have been regularly refused entry into Israel/Palestine ever since.[1]

Israel's securitization of the border has caused immense trauma to Palestinians in Jordan, including separating families and limiting their educational and job opportunities. This rather impenetrable border has also fueled the sedentarization and territorialization of Palestinians in Jordan, as their mobility westward is severely hindered.

Palestinian refugees in Jordan are living in protracted displacement, which has now spanned several generations. The large majority of Palestinians in Jordan were born in Jordan and have lived their entire lives there. Further, most have Jordanian citizenship and are integral parts of the social, political, and economic fabric of Jordan. Thus, it is not too surprising that most Palestinians in Jordan feel a strong connection to Jordan. Their senses of belonging to Jordan, however, do not minimize their strong feelings of being Palestinian and intense feelings of longing for Palestine.

The separate state-territories of Israel/Palestine and Jordan are a geopolitical reality that deeply affects Palestinian refugees' lives. Yet, the territorial divisions that seem so evident on maps or in mainstream geopolitical relations sideline the ambiguous and hybrid territorial entities that are part of Palestinian refugees' everyday lives, senses of belonging, and experiences with displacement. In other words, the state-territorial nexus is embedded in Palestinians' territorial imaginings and senses of belonging, but in ways that mix and blur with nonstate-territories. There are substantial variations in how Palestinians in Jordan imagine hybrid territories of Palestine and Jordan, as well as differences in the intensity with which Palestinians feel a sense of belonging to hybrid territories, yet they all demonstrate that clear state-territorial divisions do match their experiences with displacement.

In the first subsection below, I explain the ways that Palestinians in Jordan imagine the territories of Palestine and Jordan as being the same, but separate. I then move on to explain how daily life and the reality of living long-term in Jordan has facilitated strong senses of belonging to Jordan, all while most refugees maintain a longing for Palestine. In the third subsection, I discuss examples of Palestinians feeling a weak connection to Jordan, largely because they have felt discriminated against and marginalized. These people are living their lives in Jordan and recognize the value in living within Jordan but do not feel as though they belong there.

SAME (BUT SEPARATE)

Imaginings of Jordan and Palestine as being the same territory were occasionally expressed by Palestinians I interviewed in Jordan. A few Palestinians imagined Jordan and the whole of Historic Palestine (which covers the same geographic extent as the British Mandate of Palestine) as being one territory. For example, a female university student said that she did not differentiate between Jordan and Palestine. She felt that it was one united territory that was both her homeland and where she belonged. A Palestinian man likewise said, "Palestine and Jordan are the same; they have the same dream of being united and free." These two Palestinians were, of course, fully aware of the geopolitical separation of the two territories; and they both stressed that it was the British mandates and then Israel's expansion that had divided what they imagined as one territory and one people spanning both sides of the Jordan River.

Sameness was sometimes conceptualized through geophysical factors, like soil or a "thin" river. A Palestinian woman in her sixties talked about Palestine and Jordan sharing the same soil, and she explained that this soil was what fed them all and thus connected people on both sides of the state-territorial border. An elderly Palestinian man imagined the two territories as united and said that it was only a thin river, which could be walked across in many places, that divided Jordan and Palestine.

Other Palestinians thought of the sameness of Jordan and Palestine in terms of Jordan and the West Bank, as opposed to Jordan and the whole of Historic Palestine. This is not too surprising, since the West Bank was an official part of Jordan from 1948 to 1967.[2] Several Palestinians I met were quite nostalgic for this nineteen-year period when the West Bank and Jordan were united administratively and materially as one territorial state. During this time, they could travel back and forth across the Jordan River with relative ease. A Palestinian man in his seventies explained that during those two decades, "there was no Palestine and Jordan. It was the West Bank and the East Bank and they were one unit." A Palestinian woman in her late sixties told me that before 1967, her family would go to the West Bank for lunch to visit family and then return to Jordan on the same day. She loathed that Israel had invaded and occupied the West Bank in 1967, dividing this once united territory and creating widespread trauma and immobility for Palestinians on both sides of the border. Since 1967 and Israeli's control of the border, she has been separated from her family in Nablus (a Palestinian city in the West Bank). She explained that the distance between Amman (where she lived) and Nablus was less than the distance between Jordan's two main cities of Amman and Irbid (43 and 59 miles, respectively), but it takes her six hours to cross to Nablus because of the security checkpoints, whereas getting to Irbid takes a bit more than one hour. While she was clearly aware of the geopolitical division and had experienced the highly securitized border, her imagining of a united West Bank/Jordan remained strong. Other Palestinians I met, typically older ones, shared similar stories of how they used to be able to visit family and friends in the West Bank and return to their homes on the east side of the Jordan River on the same day. A man in his seventies said, "I wish that the relationship between Jordan and Palestine were the same as it was before 1967. They were one country, one government, and crossing borders was allowed without limitation. I could drive from Palestine to Amman directly. My friends in Palestine used to drive from Nablus to Amman as I drive throughout Jordan now and nobody will stop you."

While these Palestinians imagined Jordan and Palestine as being the same, nearly all of them also recognized the reality of their geopolitical separation. This same-but-separate status was commonly noted, and many Palestinians used metaphors to describe it. For example, a young Palestinian referred to Palestine and Jordan as "a table that was broken in half." He elaborated that "Palestine and Jordan are two fragments of something that was once whole." One man referred to Jordan and Palestine as being two sides of the same coin, as he imagined them as a singular entity, but with different sides. Familial metaphors were particularly common ways that Palestinians expressed their imaginings of sameness of Jordan and Palestine. Some referred to the relationship between Jordan and Palestine as being that of "sisters," "kin," or "twins" (see also Gandolfo 2012, 102). A middle-aged Palestinian man who did not have Jordanian citizenship referred to Palestine and Jordan as "twins," and another man called Palestine and Jordan "a family." An elderly Palestinian woman referred to Jordan as "the in-law," similar to "when a woman marries and becomes part of another family." And a Palestinian man in his forties referred to Jordan as his "loving mother."

The sentiment that Palestine and Jordan are the same, but separate, was also reflected in how people self-identified. It was very common for the Palestinians I interviewed to refer to themselves in hybrid terms as Palestinian-Jordanian or Jordanian-Palestinian. For example, an elderly married couple living in a rural area outside of Irbid hung streamers of small Palestinian flags and "We Are All Jordan" flags outside their house (figure 5.1). The woman had been displaced from Palestine when she was a baby, and the man had been born in Jordan to parents who were forcibly displaced. The husband explained that he hung the two different flags in their yard because he felt that they represented two similar things that defined him. His wife was a bit more reticent about the importance of the symbols, but she asserted to me that she was both Jordanian and Palestinian and that they were one and the same.

FIGURE 5.1. Palestinian and We Are All Jordan flags. A Palestinian Jordanian family displays these flags outside their home in northern Jordan.

BELONGING TO JORDAN BUT LONGING FOR PALESTINE

Concomitantly to imagining Jordan and Palestine as the same although separate, many Palestinians I met also stressed deep connections with and senses of belonging to Jordan. Considering that most Palestinians in Jordan were born and have lived their entire lives there, their sense of attachment to Jordan is not surprising (Mason 2007). A young Palestinian man was clear that while he loved Palestine, he felt that he belonged in Jordan, because he had lived his entire life in Jordan and had never been to Palestine. He continued that he wished things were different and that his life were in Palestine, but he acknowledged that was not his reality. A Palestinian woman in her fifties echoed that sentiment, telling me, "I love Palestine, but I know nothing about it. I grew up here. So Jordan is my home." A woman in her thirties felt that my question—where she felt she belonged—was an odd thing to ask, because as we sat in her home with her three children milling around, she thought it was entirely obvious that her life was in Jordan and thus that was where she belonged. An older man who had been born in Palestine and forcibly displaced in 1948 said that he felt Palestinian first and foremost. However, because he had spent two-thirds of his life in Jordan and worked and raised his family there, Jordan was his home and where he belonged. A woman in her sixties who had been born in Palestine but lived most her life in Jordan was clear that Jordan was her home, because, as she explained, Jordan was where she "ate and drank" every day. Senses of belonging in each of these examples stemmed from the Palestinians' daily reality, of living their lives in Jordan. Yet, they still identified as Palestinian and had intense feelings of longing for Palestine, a point that I elaborate on in the next subsection.

Palestinians in Jordan often said that compared to Palestinians who lived inside Israel and the Occupied Palestinian Territories of the Gaza Strip and the West Bank, that they had a good and comfortable life where they felt safe and secure and had many opportunities to better their lives. A young man in his early twenties explained, "Jordan gives me hope, safety, warmth, rights and duties; Palestine cannot." Another young man in his twenties, who labeled himself "Palestinian and Jordanian," referred to Jordan as where he could have a "future." He elaborated that living in Jordan was a situation that he had not chosen and that he would much prefer to live in a free Palestine, but living in Jordan was his daily reality and he made the best of it. This sentiment, that living in Jordan is not ideal but that it is a better life than Palestinians have in Israel, the Occupied Territories, and neighboring Arab states like Syria, Lebanon, and Egypt was often expressed with quite a bit of guilt and regret. Palestinians outside Jordan are commonly believed to be suffering immensely more than those in Jordan, the latter of whom have been given both safety and the opportunity to build a wealthier future. A similar sense of guilt was also conveyed by Palestinians with Jordanian citizenship toward the Palestinians in Jordan who had been displaced from the Gaza Strip in 1967 and were stateless "guests" in Jordan with limited rights.

Crucially, having a sense of belonging to Jordan did not, for the most part, dilute Palestinians' memories of or attachments to Palestine, nor their conviction of their right of return (Gabiam and Fiddian-Qasmiyeh 2017; see chapter 6). Most Palestinians who expressed a sense of belonging to Jordan also asserted that Jordan was not a replacement for Palestine. Indeed, those that imagined Jordan as their home almost always nuanced this imagining by asserting that their sense of belonging to Palestine was stronger, more meaningful, and deeper than their sense of belonging to Jordan. For example, an elderly man said, "I love Jordan, I am Jordanian, but I love Palestine more." A woman who was born and raised in Jordan and had never been to Palestine asserted that though she loved Jordan, her "soul belongs to Palestine." And a woman in her sixties who had been displaced from Palestine as a young girl said that she "loved Jordan," but she dreamed of going back to Palestine. And a mother and daughter said to me in near unison, "We are Jordanian, but Palestinian at heart." In all these examples, Palestinians were expressing that their sense of belonging to Jordan coexists with their longing for Palestine.

LIVING IN JORDAN BUT EXCLUDED

Palestinians are a part of the fabric of Jordanian society (see chapter 3). Those with citizenship also, for the most part, have political rights. Yet many Palestinians in Jordan feel like they do not belong there. One major reason for this feeling of exclusion is that some Palestinian refugees in Jordan have experienced discrimination and marginalization. Marshood (2010) refers to discrimination against Palestinians in Jordan as "obvious" (35) and argues that it is clearly exemplified through the high rates of poverty and unemployment among Palestinians. In many cases, feelings of being marginalized and/or excluded from Jordan propelled Palestinians' feelings of belonging to Palestine, all while living in Jordan.

Several Palestinians I met who had Jordanian citizenship said that they did not feel included in Jordan and that they experienced discrimination as Palestinians. Several times, my interviewees said that they felt they were treated as "second-class citizens." Three Palestinian men in their twenties spoke about the government's nation-building campaigns of "Jordan First" and "We Are All Jordan" (see chapter 3). They felt these campaigns were blunt examples of the government's attempt to Jordanize its Palestinian residents, and thus to erase their Palestinian identity. Some Palestinians told me that they felt the need to suppress their Palestinian identity in public in order to avoid discrimination. For example, a few Palestinians said that they flew the Palestinian flag or wore a black-and-white kaffiyeh (a scarf worn by men that symbolizes Palestine) but that they did so only inside their homes or in the Palestinian refugee camps. While they were proud of being Palestinian, they worried that bearing these symbols in public was too political and could lead to discrimination.

Perhaps unsurprisingly, Palestinians without citizenship were much less likely to express a sense of belonging to Jordan than those with citizenship. For the 1967 refugees (officially "displaced people") who do not have citizenship, their exclusions are codified in laws and policies. Those without citizenship are subject to limitations on starting businesses, owning land, and obtaining employment, while also incurring higher fees to attend public universities. Discrimination and exclusion was, for several of my interviewees, most notable in the context of the job market. One young Palestinian man who had been born in the Gaza refugee camp (officially named the Jerash camp) in Jordan and who did not have citizenship explained that though he had just completed a university degree in education, he could not find a job because employers were leery of his status as a noncitizen.

A Palestinian elder who is a leader in the Gaza refugee camp and who is a 1967 "displaced person" explained that even though he had spent his entire adult life in Jordan, he did not feel Jordanian. Yet he is a part of Jordan, working for the Jordanian Department of Palestinian Affairs and enmeshed in the politics and economy of Jordan. Further, while he does not feel Jordanian, he does feel a strong sense of belonging to the Gaza refugee camp, which is located just outside of Amman and is an integrated part of Jordan (see chapter 7).

Syria and Jordan

Prior to WWI, modern-day Syria, like Palestine and Jordan, was a part of the rather open provinces of the predominantly Arab area of the Ottoman Empire. With the division of Southwest Asia after WWI, Syria (which included Lebanon) was created as a French mandate. Then, with its independence from France in 1946, Syria became a state-territory that fits within the modern spatial ordering of the world.

The commonalities between the states of Syria and Jordan, as well as between Syrians and Jordanians, are significant and numerous, but the relationship is not as deeply entwined as that between Palestine and Jordan. Syria and Jordan were not territorially united in the twentieth century, as Jordan and the West Bank were; and Syrians' mass displacement into Jordan has been much more recent. Nevertheless, there are many historical, cultural, economic, and geopolitical interconnections that have deeply affected Syrian refugees' experiences with displacement and their senses of belonging to both Syria and Jordan. For example, visa-free passage between Jordan and Syria has facilitated labor movements, trade, familial visits, and vacations for many decades (though Jordan recently stopped this practice). Prior to the Syrian war, extended families lived on both sides of the border, and marriages between Syrians and Jordanians were not uncommon. Economic, business, and labor relations between Syria and Jordan were numerous and were particularly evident in northern Jordan, in and around Mafraq, Ramtha, and Irbid, where many Syrians had well-established business ties and jobs. During the March 2011 Arab uprising protests in Jordan, Jordanian crowds in the border town of Ramtha invoked these long-standing cross-border connections, chanting "One and one, Dara'a and Ramtha are one!" (Schwedler 2022, 165).

The decades of cross-border connections between Jordan and Syria, buttressed by older historical and cultural interconnections (see chapter 4), have affected Syrian refugees' senses of belonging, as well as their decisions about migrating to and staying in Jordan. In the next subsection, I explain the different ways that interconnections between Syria and Jordan (and Syrians and Jordanians) that predated the Syrian war have helped some Syrians to settle and feel as though they belong in Jordan. Then I explain how daily life since the war—the reality of living long-term in Jordan as guests and refugees—has cultivated new interconnections and senses of belonging in Jordan. However, while some Syrians feel a sense of inclusion in Jordan, other Syrian refugees—whether due to prewar factors or to their daily experiences in Jordan—have experienced discrimination and exclusion that has created or deepened cleavages of difference. Indeed, many Syrians do not feel any sense of attachment or belonging to Jordan and believe the Syrian war and their displacement into Jordan has exacerbated preexisting differences between the two states.

LONG-STANDING INTERCONNECTIONS AND INCLUSION

Many Syrians I interviewed felt that Jordan was the most reasonable place for them to seek refuge. Proximity, particularly for Syrians who had lived in the south of Syria, was one pragmatic reason for them to have fled to Jordan, but so too were the preexisting similarities in their cultural practices, imaginings of *Bilad al-Sham*, and economic ties.

One woman told me that she "blended" easily in Jordan and that she had made many Jordanian friends. She believed that her inclusion in Jordan was because she shared many customs and traditions with Jordanians. Syrian and Jordanian cultures "were already similar," another Syrian woman told me, and now that they were living among one another, they were learning even more about each other's traditions and sharing their practices. Several Syrians told me that they would not feel as comfortable in any other state, including other Arab states, because no other state was as culturally close to Syria as Jordan.

Twentieth-century geopolitics was not as important a factor in creating interconnections for Syrians as it was for Palestinians. But a few Syrians mentioned that Jordan and Syria had once been a part of *Bilad al-Sham* and that this older territorial unity still had meaning for them. For example, a Syrian man in his forties and a Syrian woman in her late twenties both referred to *Bilad al-Sham* as their "homeland" and felt that they were still in this homeland even though they were displaced in Jordan.

Decades of shared economic and employment opportunities have facilitated interconnections between Syria and Jordan. A few Syrians I interviewed had family members who had worked in Jordan. One Syrian man I met had worked in construction in Jordan for twenty years before the war erupted. When the war broke out, he was already in Jordan working, so he got his wife and son across the border to join him in Irbid. This family was able to cope with their forced displacement, including finding an apartment and maintaining an income, with relative ease because of the man's decades of work building relationships and connections with Jordanians.

These examples of preexisting cultural similarities, territorial imagining of *Bilad al-Sham*, and economic ties have influenced Syrians' decisions to come to Jordan and to settle and rebuild their lives there. The long-standing connections between Syria and Jordan were so valued by some Syrians that they rejected the opportunity to be resettled in a third country, which is one of three durable solutions. I interviewed two different Syrians whose families had both declined resettlement offers—one to Germany and the other to the US—because they wanted to stay in Jordan, where they felt connected and comfortable. A Syrian man in his thirties whose family had declined resettlement explained, "My wife got the call to go to Germany, but we said no. . . . We wanted to be in Jordan . . . because the community in Jordan is so similar to Syria." In addition to prewar interconnections, as I discuss in the next section, the experiences of living in Jordan and interacting with Jordanians due to the war has created new or deeper interconnections.

NEW INTERCONNECTIONS THROUGH DAILY LIFE

As years and years have passed, many displaced Syrians have become dismayed by their prolonged stay in Jordan and worry that they may not return to Syria. Most Syrians are adamant about wanting to return home (see chapter 6), but after years of coping and adapting in Jordan, Syrian refugees, like Palestinian refugees, have remade their homes and lives in Jordan out of necessity. As time has passed, many have come to feel comfortable and happy in Jordan. They have made new friends, built new communities, and developed a sense of belonging. One woman explained that while she had been afraid at first as a refugee in Jordan, she had become comfortable and felt as though she now belonged there. And a Syrian man in his sixties told me, "Jordan opened their borders and gave us life and security," enabling him and his family to make Jordan what he referred to as their "second home."

Living in Jordan, for as many as eleven years, has created new connections between countless Syrians and Jordanians, whether as neighbors, friends, or colleagues. Many Syrians felt that Jordanians were friendly, open, and generous toward them. A Yarmouk University study (Basset and al Mooney 2016) on how Jordanians felt about Syrian refugees provides statistical data that helps explain why Syrians often felt welcomed. The study found that 77 percent of Jordanians deal with Syrian refugees as fellow Arabs or "as a brother," whereas 19.8 percent consider the refugees Syrian, and only 3.3 percent deal with Syrian refugees as foreigners.

Many Syrians also felt that the Jordanian government was generous and honorable, and this cultivated a feeling of being welcomed in Jordan, which, in turn, has helped some Syrians to feel as though they belong. Most appreciated the aid and services that had been provided by the state, even if scant. A few Syrians told me that they felt fortunate to have Jordan's support particularly because they knew that Jordan's economy was struggling. A married couple I interviewed together heralded Jordan as "saving" Syrians. To show her appreciation toward Jordan, a young Syrian woman living in Za'atari used dried, colored beans to make a "Jordan First" decorative wall hanging for one of the schools in the camp (figure 5.2). As discussed in chapter 3, the symbol asserts the primacy of Jordan and Jordanian cohesiveness. When I asked her why she made this, something that symbolically excludes her and other Syrians, she explained that it was a gift to show her thanks. She elaborated, "Jordan stood by

FIGURE 5.2. Wall decoration reading "Jordan First." This was made by a Syrian refugee in gratitude for Jordan's help for Syrians.

us, they didn't forget about us, and they helped us during our time of distress." Those actions made her feel welcomed, even though she is not Jordanian.

EXCLUSION IN JORDAN

While some Syrians felt included in Jordan, others felt excluded and discriminated against by numerous different people and actors. Syrians often noted that they felt looked down upon, disrespected, and mistreated. One woman I met referred to her maltreatment in Jordan as a "second trauma." Such bad experiences, of course, have made it difficult for some Syrians to form senses of belonging to Jordan. Indeed, several Syrians told me that they did not feel as though they belonged in Jordan, regardless of any preexisting territorial, economic, cultural, or historical interconnections.

In general, both the Jordanian government and Jordanian citizens welcomed Syrian refugees at first, but over time some tensions emerged and their welcome became less warm (Yassin-Kassab and Al-Shami 2016, 155, 157; Chatty 2018). The mass displacement of Syrians into Jordan has created new problems and has exacerbated existing ones—like increased rents, decreased job availability, and increased cost of living—for the already struggling state. One Syrian woman told me that the general sentiment she feels on the street from Jordanians is that "Syrians have made everything expensive in Jordan and thus are destroying the country."

For some Syrians, preexisting connections between Syria and Jordan did not matter. Being a refugee effectively erased such connections. One Syrian woman explained that Syrians suffer in Jordan "not because we're Syrians, but because we're refugees." Another Syrian woman was devastated at how relations had changed between Syrians and Jordanians due to Syrians' displacement. Prior to the war, her family had been close with a Jordanian family, and she had hosted this family in Syria years ago. However, when she arrived in Jordan with her family as refugees and reached out to the Jordanian family, that family refused to even see them.

Taxi drivers were commonly singled out as the people who caused Syrians the most difficulties—sometimes refusing service to Syrians or overcharging them for rides. Landlords who exploited the housing shortage and increased their rents were loathed by many Syrians (and Jordanians). One Syrian woman referred to landlords as "heartless." She, like many other Syrians, was baffled and hurt that some landlords took advantage of Syrian refugees who were in such precarious situations. Likewise, some employers exploited Syrian laborers by underpaying them or forcing them into dangerous working conditions.

Schools are a problematic space of exclusion. The Jordanian government has provided free public education for Syrian youths, but parents commonly told me that they worried for their children, who experienced microaggressions and insults from some teachers and students. A few mothers explained to me that their children had stopped going to school because they were so uncomfortable in the Jordanian schools.

Some Syrians told me that strangers on the street or in stores occasionally harassed or cursed them. A mother shared with me an experience she had at a public park, when a young Jordanian boy refused to share a swing with her son. The Jordanian boy yelled to her son, "The swing is ours, not yours!" Reflecting on this incident, she wondered if this little boy had viewed her as a human being. The comments that some Syrians heard on the streets were so difficult to handle that they would avoid public spaces and/or try to hide the fact that they were refugees. Two women told me that they used a Jordanian dialect when communicating with Jordanians, to blend in a bit more and divert any harassment. Women's experiences on the streets and in their communities could differ quite starkly from men's. Women, particularly single or widowed women, were at times fearful that male strangers might sexually harass or threaten them.

Several Syrians considered the government to be discriminatory and oppressive, particularly because of its policies like encampment and *kefala*. A few Syrians said they felt like they were being constantly watched by state officials. A couple of Syrians said that the border guards and soldiers they interacted with were hostile, rude, or callous.

These negative experiences have resulted in some Syrians feeling excluded from Jordan and different from Jordanians. One man told me that Jordan could never feel like his home because of all the maltreatment and discrimination that he had experienced. Yet the palpable experiences and feelings of being discriminated against coexist, for the most part, with positive experiences and senses of feeling included. A few Syrians recognized that discrimination happened but felt that it was not ubiquitous. For example, a Syrian man explained that while some Jordanians looked down upon, insulted, or demeaned Syrian refugees, most did not. Another Syrian man echoed this idea, saying, "Some people look down on Syrians, but these are very few." And one woman expressed that she was confused and saddened that some Jordanians could treat Syrians so poorly, but she was simultaneously appreciative of all that Jordan had done for Syrian refugees and told me that if she could not live in Syria, then Jordan was where she wanted to stay.

In Summary

Seven decades of displacement, living and raising families in Jordan, and Jordan's control of the West Bank from 1948 to 1967 have led to Palestinian refugees in Jordan having varied senses of belonging to the mixture of both Jordan and Palestine, as well as cultivating hybrid Palestinian-Jordanian identities. A few Palestinians I interviewed imagined Palestine and Jordan as one united territory where they belonged, whereas most others recognized interconnections but difference, too. Regardless of how much Jordan and Palestine blurred and mixed in their imaginings, the reality of the highly securitized border that divides the state-territories of Israel/Palestine and Jordan was never denied. Some Palestinians I interviewed expressed great love for Jordan and a strong sense of being included there, while also embracing their Palestinian identity and asserting that they loved Palestine more. Others, both citizens and stateless Palestinians, felt marginalized and excluded in Jordan, while recognizing that they were a part of the Jordanian state and lived each day in Jordan. Despite the depth or intensity of their senses of belonging and interconnection, most Palestinians recognize the reality of their lives in Jordan and mix their sense of belonging to Jordan with their longing for Palestine. Crucially, senses of belonging to Jordan are not in competition with belonging to Palestine but coexist and often blend together (Massad 2001, 263; Culcasi 2016). Yet for some Palestinians, as I discuss in the next chapter, resolute feelings of being Palestinian and maintaining a deep sense of belonging to Palestine overshadow their connections with Jordan.

Syrian refugees' interconnections with Jordan are not—for the most part—as long-standing or as strong as those of Palestinians. This is largely because Syrian refugees are more recent arrivals, because they have not been granted citizenship or permanent residency, because they have an independent state to which they can (at least theoretically) return to as citizens, and because modern-day Jordan and Syria were never united as Jordan was with the West Bank. Nevertheless, drawing from preexisting interconnections, including territorial openness, similar cultural practices, and cross-border employment and business, many Syrians feel comfortable and included in Jordan. Further, for many refugees, rebuilding their lives in Jordan and daily interaction with Jordanians have created new and/or stronger senses of connection between Syrians and Jordanians. Yet for other refugees, those preexisting and/or new interconnections matter little. These Syrians' feelings of being mistreated or discriminated against have weakened a sense of connection with and belonging to Jordan.

The numerous ways that Palestinian and Syrian refugees imagine interconnections with Jordan, as well as their senses of inclusion and exclusion there, have had significant impacts on their decisions about movement and settling. This, in return, has immense influence on how they rebuild and live their lives. The territorial imaginings and senses of belonging that I highlighted in this chapter neither fully conform to nor reject the dominant world ordering of states and the ideal of each person belonging to one state. Rather, they confirm that people often imagine territories and their senses of belonging in hybrid ways.

6

The State-Territory Nexus

The state-territory nexus is integral to the political ordering of the world, the laws and practices of the IRR, and many of Jordan's policies, as discussed in chapters 2 and 3. Further, state-territories are also formative of refugees' imaginings and senses of belonging. Indeed, it is entirely common for displaced people not only to maintain connections to the states from where they were displaced but also to feel intense senses of loss, longing, and belonging with respect to those state-territories.

In this chapter, I focus on the varied ways that Palestine and Syria, as a quasi state and an independent state, respectively, are central to refugees' territorial imaginings and senses of belonging. In the sections below, I discuss the different ways that displaced Palestinians and Syrians imagine Palestine and Syria as variously (1) conventional, as a bounded, discrete territory under the control of a state, (2) ambiguous and amorphous, as definitively somewhere else but also unlocatable, and (3) abstract, as feelings and embodiments of the conventional territories. Below, I provide some necessary context into the production of Palestine and Syria as state-territories and then delve into refugees' imaginings of each. Throughout this chapter, I show the centrality of state-territories in Palestinian and Syrian refugees' imaginings, senses of belonging, and experiences of forced displacement.

Palestine as a (Quasi) State-Territory

There is, arguably, no territory in the world more multifaceted and contentious than Palestine.[1] Palestine is not an independent state-territory in today's world order. It is more aptly described as a quasi state, which is a recognized political and territorial entity that lacks sovereignty, autonomy, and independence (Natali 2010, xxi). Within the UN, as of 2012, Palestine has had that status as a "non-member observer state," meaning that it has a standing invitation to participate in General Assembly sessions. Further, as of 2020, a large majority of UN member states recognize Palestine (138 of 193). However, some of the most powerful states and some permanent Security Council members, such as the US, Canada, many western European states, Australia, and Japan, do not recognize Palestine as either a state or a quasi state.

During the British Mandate of Palestine from 1923 until 1948, Palestine became an internationally recognized, clearly defined territorial entity under British control.[2] Prior to then, the area that would become the Mandate of Palestine was part of the Arab-majority provinces of the Ottoman Empire. There was no Palestine vilayet or specific administrative unit labeled "Palestine" under the Ottoman Empire. Instead, the area that became the Mandate of Palestine in 1923 closely resembled the Ottoman administrative district of Jerusalem and the southern half of the Beirut district. Nevertheless, as many historians have asserted (Khalidi 1997, 28–29, 153–54; Sluglett 2010, 43; Chatty 2010, 180–81), the idea of Palestine as its own distinct entity did exist under the Ottomans.

In May 1948, the British ended the mandate and withdrew from the territory. Immediately afterward, David Ben Gurion, the prominent Zionist leader, declared the existence of the state of Israel. This announcement sparked the first Arab-Israeli war, which led to nearly a year of military combat, the displacement of approximately 750,000 Palestinians, the destruction of around four hundred Palestinian villages (Davis 2011, 3), and numerous territorial gains for Israel. The armistice that followed combat divided the former British Mandate of Palestine into the state of Israel and the noncontiguous Palestinian territories of the Gaza Strip and the West Bank. This war is referred to as "the Catastrophe" (*al-Nakba*) by Arabs.

Since the 1948 war, numerous conflicts over these territories have continued. Most significantly, in 1967, another war broke out and the Israeli military invaded and occupied the Palestinian territories of the West Bank and the Gaza Strip (as well as the Golan Heights of Syria and the Sinai Peninsula of Egypt). Since 1967, Israel has maintained power and dominance over all of the former Mandate of Palestine and has systematically oppressed Palestinians in both Israel and the Occupied Territories of the Gaza Strip and the West Bank. Palestinians have some limited control in small enclaves within the Occupied Territories and have a quasi-functioning political structure in both. Yet Israel maintains strict control of all the borders, including controlling and hindering the movement of goods and people in and out of the occupied Palestinian territories. Thus, while there are tiny, segmented territories in which

Palestinians ostensibly have control, Palestine does not function as an independent territory but, instead, as a quasi state.

The status of Palestine as a quasi state does not negate or weaken Palestinians' senses of belonging to Palestine, their commitment to Palestine's liberation from Israel, or their pursuit of the legal right to return, which is enshrined in UN Resolutions 141 and 242. This commitment to continue the struggle to return to Palestine (referred to as *sumud*) has been fueled through their collective experiences of suffering and injustice (Gabiam and Fiddian-Qasmiyeh 2017). Their protracted displacement, their decades of experiences with conflict and oppression, and the lack of global recognition of Palestine as a fully independent state-territory have for many Palestinians redefined and intensified Palestinian identity and their struggle to liberate Palestine. As one young Palestinian woman in Jordan told me, "the more Palestinians suffer at the hands of Israel, the more Palestinians fight and the stronger their commitment to Palestine becomes."

Palestinian refugees in Jordan (and elsewhere) imagine Palestine in many ways, as I discuss in the next three sections. Some of their territorial imaginings fit squarely within the world order of discrete territorial states. Others do not fit well into the modern world order, as they have ambiguous and amorphous extents, but they remain highly territorial. Finally, some see Palestine as an abstraction, as a territory that is not locatable or demarcated but nevertheless exists in their minds, hearts, and bodies. Even while there are substantial variations of imaginings of Palestine, all the examples I discuss below invoke connotations to Palestine as a (quasi) state-territory.

CONVENTIONAL PALESTINE

Extending in a diamond-like shape from the Jordan River in the east to the Mediterranean Sea in the west, from the Negev Desert in the south to the to the southern foothills of the Lebanese mountains in the north (see figure 3.1), maps demarcating a discrete, bounded territory of Palestine are commonplace in many Palestinian communities, businesses, and homes. This mapped territory is often found framed and hung on the wall, worn as necklace charms, dangling from key chains, gracing political posters, and posted on social media (Culcasi 2016). Indeed, many of the Palestinian refugees I interviewed in Jordan showed me maps of Palestine that they owned and used. While the decorative, symbolic, and artistic elements of the maps certainly varied, they all represented Palestine as a bounded, discrete territory that fit within the conventional imaginings of the state-territory nexus.

A few of the Palestinians who shared with me their maps noted that these served as daily reminders of their struggle for liberation. For example, a young Palestinian man who lived in the Gaza refugee camp (again, officially named the Jerash camp) walked me to a graffitied outline of Palestine on a concrete wall in the camp. It was only an outline in black and had no artistic embellishments, but nevertheless, this simple image of the extent and borders of Palestine on a gray wall was an important daily reminder to this man of his lost homeland and right to return.

It is quite common for outlines of the shape of Palestine to be juxtaposed with other symbols of Palestine and Palestinian resistance, yielding highly symbolic map-logos. A young woman living in al-Husn refugee camp embroidered a map of Palestine, using the symbolic green, red, and black colors of the Arab and Palestinian flags to create the territorial shape of Palestine. Underneath the map was a series of Palestinian flags, and written across it was the phrase "Your love, my homeland" (figure 6.1). Many other Palestinians shared with me maps and other visual reminders of Palestine. Some used remnants of olive trees from Palestine as a base on which they displayed their map. Many included the place names of villages that had been destroyed, some were decorated with the black-and-white kaffiyeh, many had the Palestinian

FIGURE 6.1. Embroidered map of Palestine. This was made by a Palestinian woman and displayed in her home in the al-Husn refugee camp.

flag, and some included images of the Dome of the Rock. Regardless of the use of symbolic decorations, each map and map-logo demarcated the existence of a discrete, bounded territory that is very real in Palestinians' imaginings, even if it does not exist as an internationally recognized independent state in the modern political ordering of the world.

The clear demarcation of Palestine in maps and map-logos and Palestinians' imaginings of Palestine as a bounded, discrete territory are important reminders of their lost territory and a symbolic challenge to the erasure of an independent Palestine at the hands of Israel. Yet these maps and the territorial imaginings of Palestine with these clear borders are also a recreation of the imperial territorial borders of the British Mandate, as well as the Western-based state-territorial ordering of the world, from which Palestine is greatly excluded.

AMBIGUOUS/AMORPHOUS PALESTINE

Discussing maps of Palestine and its precise territorial extent was a common topic during my interviews. Yet it was likewise very common for Palestinians to speak about Palestine as an ambiguous territorial entity. Many Palestinians referred to Palestine as their homeland from which they had been excluded for many decades. Palestinians commonly mentioned the Jordan River as Palestine's eastern border, but the extent of Palestine westward, as well as its extent in a north/south direction, was rarely mentioned as a matter of concern or an important geographical reference. In such discussions about the shape or location of Palestine, Palestinians often referred to Palestine as being elsewhere, somewhere they were not, and somewhere they were excluded from entering. Palestine's precise location did not matter much in these contexts. It certainly existed across the Jordan River, but its shape was amorphous, and its other borders were ambiguous.

In order to initiate discussions about territorial imaginings, I often brought maps with me to interviews, including an unlabeled map of Southwest Asia. This outline map (figure 6.2) uses the common internationally recognized borders that distinguish Israel from the Occupied Territories of Gaza and the West Bank. This map differs dramatically from that of Historic Palestine, as it recognizes Israel and fragments Palestine into two discrete, noncontiguous territories. Upon seeing this map, an older Palestinian man I was interviewing quickly corrected the map to match his imagining of Palestine, which mirrored that of Historic Palestine. He used a red marker to circle his image of Palestine and a green pen to cross out the border separating the West Bank and Jerusalem from the rest of Historical Palestine. But his decisive

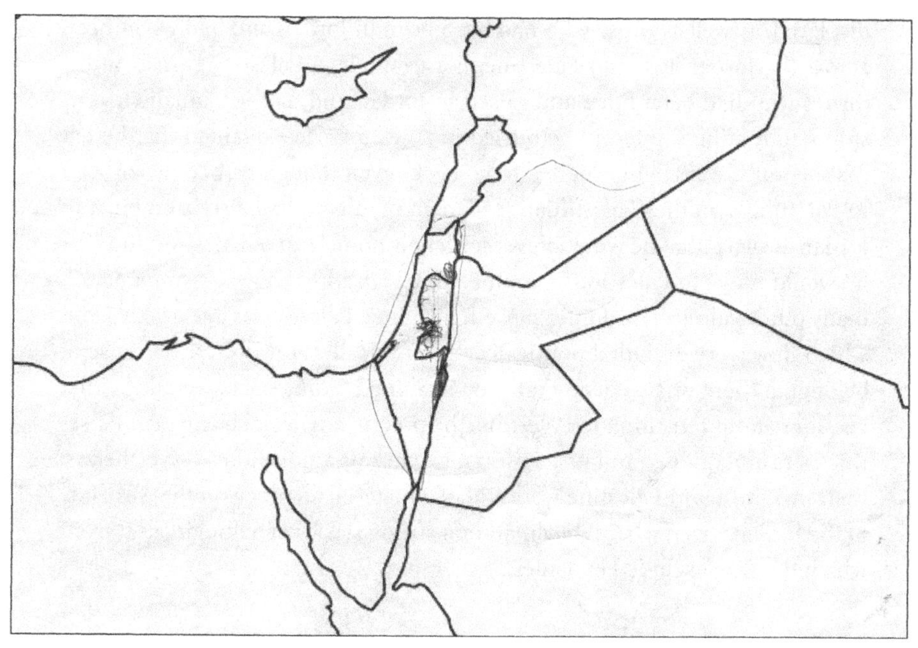

FIGURE 6.2. Southwest Asia outline map. This outline map, which I used during many of the interviews, was redrawn by a Palestinian man to match his territorial imagining of Palestine.

and clear redrawing of the map was not the norm. Most of the twenty-four Palestinians with whom I shared this map looked at it with quite a lot of confusion. The West Bank, clearly demarcated as its own separate territorial entity, was an unfamiliar cartographic territory for most Palestinians. Several people thought the demarcation of the West Bank was the Dead Sea (which is to its south) or Lake Tiberias/Sea of Galilee (which is to its north). Many of the Palestinians I interviewed spoke articulately and knowingly of the West Bank, but they did not imagine it as its own discrete and bounded territory, and they did not recognize it on this map. In other words, this delineation that is common in many places across the globe was quite unfamiliar to these Palestinians, likely because it did not fit their imaginings of Palestine, nor did it match the commonplace maps and map-logos of Historic Palestine as a discrete, bounded territory.

Most Palestinian refugees have a strong desire to return to Palestine and they commonly spoke of their longing to go *back*. Many asserted that they would return permanently if Palestine were ever free of Israel, and a few noted that they wanted to be buried there as well. This desire or dream to "go back" was expressed by Palestinians who themselves had been forcibly displaced, as well as those who had been born in Jordan and had never been across the Jordan River to Palestine. For example, a Palestinian man in his forties who had been born and raised in Jordan and had Jordanian citizenship said that he wanted to be buried in Palestine. He explained that he felt that he was always being "pulled to go back" even though he had never been to Palestine. Another Palestinian Jordanian man who had also been born in Jordan asserted that he would never forget his homeland of Palestine and that he would move to Palestine as soon as Israel withdrew. These two men, and many other Palestinians I interviewed, imagined Palestine as a territory from which they were excluded politically and physically. However, their sense of belonging there was so deep that they imagined "going back." In all the discussions about returning to Palestine, the precise territorial borders of Palestine were ambiguous, and its territorial extent was amorphous. Nevertheless, their imaginings of Palestine were still of a material territory on the west side of the Jordan River, where they maintain strong senses of belonging but from which they are excluded by Israel.

ABSTRACT PALESTINE

Edward Said (1992) famously wrote that Palestine is, in part, a "consciousness." Other scholars have likewise referred to it as abstract and embodied (de Vet 2007; Johnson and Shehadeh 2013; Barakat 2013). And for Palestinian refugees in Jordan, they too often described Palestine in quite emotional and embodied ways. Imaginings of Palestine as abstract still invoke and signify a territory, but it is a territory that is defined much more through emotions than through borders.

Several of my interviewees spoke of Palestine as being embodied within them. They explained to me that Palestine was not on the map, but instead it was a part of them, in their hearts and in their blood. A man in his seventies told me, "It is not about borders," and then clarified that "Palestine is in my heart." Another older man told me that Palestine "is engraved in my heart," and a young woman said "Palestine lives in me." A Palestinian man in his forties waved his hand dismissively at a map of Historic Palestine that I showed him and then continued to move his arms around as he told me that "Palestine is here, there, and everywhere."

It was also quite common for Palestinians to invoke abstract territorial imaginings through symbols of Palestine. The kaffiyeh, as noted above, is an important symbol of Palestine that at times adorns people's bodies and is used as a decorative element on many symbolic objects. A Palestinian woman in her early twenties told me that she always carries her black-and-white kaffiyeh in her bag to remind her of Palestine and of their right to return. Handala, a cartoon figure of a boy created by celebrated Palestinian artist Naji al-Ali in 1969, is another common symbol of Palestinian resistance. Handala appears in many mediums today, like the key chain shown in figure 6.3 which depicts him alongside a map that is filled with a representation of the Palestinian flag. Handala, like the kaffiyeh, is not an inherently territorial image or symbol that directly connotes a conventional idea of a discrete, bounded territory, but both nevertheless invoke powerful memories and imaginings of Palestinian territory, from which most Palestinians have been excluded for seven decades.

Palestine is also remembered and imagined through particular sites in Palestine or through materials that originated from Palestine. It is not uncommon for Palestinians to keep photographs or images of Jerusalem, the Dome of the Rock, and al-Asaqa mosque in their homes or on their social media pages. These images work as reminders not only of these exact sites but also of Palestine more broadly. Palestine is also abstracted from material objects that originated in Palestine. Remnants of olive trees that once grew in

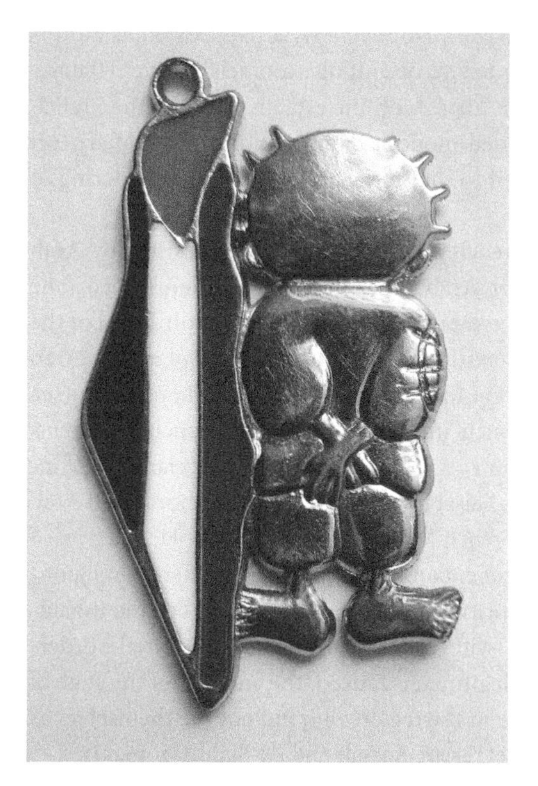

FIGURE 6.3. The Palestinian cartoon character Handala with a map/flag logo. This key chain bears three different symbols of Palestine, one of which is the highly territorial symbol of a map.

Palestine are commonly repurposed into plaques on which different symbols of Palestine, including maps, are displayed.

Abstract, ambiguous, and clearly defined imaginings of Palestine are all commonplace, and they work in tandem to keep memories and the struggle to return alive. A Palestinian man explained to me that though he hung a map of Palestine on his wall at home, a map that recreated the borders of Historic Palestine, those clearly demarcated borders did not define Palestine for him. Palestine was, instead, a vague place on the other side of the Jordan River, the place from where his family had been displaced. He continued that Palestine was also a place in his heart, a place that could not be mapped, and a place he dreamed of returning to. As famed Palestinian poet Mahmoud Darwish (2010, 113–14) wrote to his fellow Palestinians in 1973, even though Palestine was under military control, "Palestine remains your homeland, be it a map, a massacre, a land or an idea."

Syria as a State-Territory

Syria has existed in different territorial forms for several centuries. At the end of the nineteenth century, the Ottomans established the Syria vilayet (also referred to as the Damascus vilayet). However, Syria had existed as an unofficial, ambiguous region long before becoming an official vilayet (Pitcher 1968; Held and Cummings 290). Then, with the establishment of the French Mandate of Syria (which included modern-day Lebanon) in the aftermath of World War I, Syria was added to the state-territorial political ordering of the world.

The establishment of the French Mandate of Syria was resisted by Arab elites and leaders who vied for Arab independence. In March 1920, one month before the mandate was created, Arab elites, led by Emir Faisal of the Hashemite family, declared the existence and independence of a new "Arab Kingdom," which was centered on the broader region of *Bilad al-Sham* (see chapter 4) as opposed to the much smaller Syria vilayet. French politicians and armies responded quickly and aggressively to the Arab declaration of independence. They swiftly exiled Faisal to Iraq, abolished the short-lived Arab Kingdom, and established the French Mandate of Syria in its place.

The French Mandate of Syria encompassed people of different religious and ethnic groups, but the majority were Arab and Muslim. For the inhabitants, this new mandated state, with its Western-defined territorial borders and imperial occupiers, lacked legitimacy. Indeed, the creation of the state of Syria was devastating for people in the region who did not see themselves as part of this new Syrian territory (Yassin-Kassab and Al-Shami, 2016, 5).

In 1946, the French Mandate of Syria ended, and Syria gained its independence.[3] As with many postcolonial states (Jordan is an exception), Syria's transition to independence has been wrought with instability and conflict. Notably, as discussed in chapter 4, in 1958, Syria united with Egypt to create the new territorial state of the United Arab Republic (UAR). This new territorial state was noncontiguous but still functioned like a singular territorial state, with Egypt's Gamal Abdel Nasser as president and with full recognition and a seat in the UN. The UAR lasted until a coup in Syria in 1961 facilitated its demise. Syria, officially known as the Syrian Arab Republic since the end of the UAR, was then engulfed in a decade of political upheaval until another coup in 1971 brought Hafez al-Assad to power.[4] Assad brought political stability to Syria, but he did so by reigning as an authoritative and repressive ruler. With his death in 2000, his son Bashar al-Assad assumed the presidency. Though the younger Assad began his rule with the promise of liberalizing Syria, he soon retreated back into the authoritative rule that had defined the past three decades under his father. Thus, since the early 1970s, Syria has functioned as an authoritative state with power centralized in the hands of a father-and-son presidency.

In March 2011, as the so-called Arab Spring spread from Tunisia to Egypt to other states across the globe, Syrians mobilized and revolted against four decades of oppressive rule under the Assad regime. Unlike the leaders in Egypt and Tunisia, Assad refused to step down and instead retaliated against protesters with extraordinary violence. Since March 2011, Syria has been consumed in a brutal war that has led to the death of approximately 380,000 people and the displacement (internal and international) of approximately thirteen million Syrians (AFP 2020). Syria has been devastated by war, but it remains an independent, internationally recognized state. Its borders are also relatively uncontested, save those of the Golan Heights, which Israel occupied in 1967 and annexed in 1981.

AMBIGUOUS SYRIA

Syrian refugees in Jordan did not use or display maps of a clearly demarcated Syrian territory as frequently as Palestinians used maps of Palestine. Indeed, I observed only a few maps and map-logos of Syria during my interviews with Syrian refugees (those few maps were painted on caravans in the Za'atari camp alongside other symbols of Syria). This distinction is likely because Syria's territorial existence is not threatened by another state and thus there is not a great need to invoke the image and memory of Syria as a clearly defined territorial state. Nevertheless, Syrians still imagined Syria in territorial ways, but these imaginings were of an ambiguous Syria, which was to the north, where they belonged, and where they were not.

Many Syrians considered Jordan to be similar to Syria and appreciated Jordan's assistance (see chapter 5) but did not want to make Jordan their permanent home. Syrians expressed a near-unanimous desire to return to Syria (only a few Syrians entertained ideas of moving elsewhere permanently—typically Canada or the UK). They often spoke of returning "home," which included specific dwellings like their houses and their gardens. The villages and cities where they had resided were also commonly mentioned as the "home" to which they wished to return. While these nonstate-territories—a house, garden, village, or city—were part of their imaginings of return, most Syrians concomitantly stressed that it was *Syria* that they wanted to return to and that *Syria* was their home and homeland. Indeed, for so many of them, the desire to leave Jordan and return to Syria was intense. Two women, good friends who had both lived in the Syrian city of Dera'a, specified that it was Syria, not Dera'a, that they dreamed of returning to. They said, "Our minds and hearts are in Syria." Many Syrians expressed the sentiment that Syria was the first, only, and last place they would ever want to live. A Syrian man in his late thirties explained that he was always thinking of his return to Syria, as opposed to a certain town or city. A Syrian woman told me that she was "counting the seconds" until her return to Syria. When I asked Syrians "Where do you want to live?" or "Where is your home?" (which were standard questions for all my interviews), they typically responded emphatically that it was Syria (*Suriya*, or *al-Sham*). Alternatively, they would laugh at my question because the answer "Syria" was so obvious that it was a foolish question to even ask.

Syria was commonly spoken of in nostalgic ways. It was imagined as a place of great beauty, one to which no other place in the world compared. A Syrian man in his forties told me that "the ground in Syria is better than a castle here in Jordan." A Syrian woman said that thinking of Syria gave her goosebumps, because her love of Syria was so deep and so emotional. Another woman in her early forties said, "All I need to do is think about Syria and my heart starts pounding, telling me to go back." Some Syrians mentioned that they would remain "loyal to" or would "never abandon" Syria, even though it was being torn apart by war. A few of them expressed their extreme worry that the war would destroy Syria forever, that a united Syria would cease to exist, and that it would be divided into different sectarian territories. Their intense senses of belonging to Syria were commonly expressed in the sentiment that Syrians *would* rebuild and *would* reunite Syria. Two women echoed each other, resolutely stating, "Syrians are strong and will rebuild Syria."

Their strong senses of belonging to Syria, of wanting to return, and their nostalgic imaginings all helped recreate the idea of Syria as a territory that is elsewhere, but its precise delineations and borders were not central to their imaginings of returning to Syria. Cartographic representations, map-logos, and borders were not part of our conversations because, unlike Palestine, Syria's territorial existence is internationally recognized. The lack of discussion of borders or use of maps does not, however, question the existence of Syria as a modern state-territory. Instead, Syrian refugees imagined returning to Syria, which was a territory north of Jordan, and where they belonged.

ABSTRACT SYRIA

Non-cartographic symbols of Syria were quite commonplace in Syrians' dwelling spaces and on their bodies. The flag of the Syrian opposition, which has a design similar to other Arab states' flags, was particularly prevalent (Syrian refugees are often aligned with oppositional forces rather than the Assad regime). The opposition flag was occasionally hoisted and flying, as one would expect with a flag, but it was more commonly appropriated into non-flag forms. For example, the image of the opposition flag was painted on caravans in the Za'atari refugee camp and displayed on T-shirts, water jugs, vases, bracelets, key chains, bandannas, and scarves. Such opposition symbols could be found and purchased quite readily in shops and vendors lining the streets in Irbid and Amman. A young woman living in Mafraq, a small town close to the Za'atari refugee camp, wore a flag-logo bracelet (figure 6.4) on her wrist, hidden under her sleeve. She explained to me that it reminded her of the purpose of the war (that of revolution) and that reminder helped her to cope with the struggles of displacement. It was also a daily reminder for her to keep up her hope of Syria being liberated and of returning home.

Similarly, a woman in Irbid who knitted, sewed, and partook of other handicrafts to earn a small income was just about to finish knitting a baby's sweater to sell when I met her. She had designed the sweater as the Syrian opposition flag, with green, white, and black stripes. The sweater was almost complete, save the three stars that she was about to add to the white band. She was proud that she could make a bit of money off the sale of this sweater and happy that a baby could wear this symbol that evoked both love and hope for Syria.

The idea of Syria was also occasionally expressed in embodied and biological terms. A Syrian man in his early thirties was dismayed with his six years of displacement and was considering resettlement in another country. But he longed for Syria because, as he explained, "Syria is in my blood." His

FIGURE 6.4. Syrian opposition flag bracelet. This symbol of revolutionary Syria is quite commonplace among Syrian refugees in Jordan; here is it worn by a young Syrian woman.

imagining of Syria was simultaneously about the physicality of Syria as a place that he wanted to return to and a more abstract and embodied imagining of Syria as being part of his body. Likewise, a young woman imagined Syria as concurrently abstract and physical. She said, "Syria is still in our hearts, it is still our homeland, and there is no way to give it up."

The use and co-opting of Syrian symbols and the sentiment that Syria is a part of the refugees' bodies does not rely on defining a discrete, bounded territory. Instead, these are abstract ways that feelings and memories of Syria as their homeland are invoked, as the territory where they belong but have been forced to flee.

IN SUMMARY

Syrian and Palestinian refugees in Jordan long to return to many different places, including specific villages, towns, cities, and dwellings; as well as pre- and anti-imperial territories (see chapter 4) and hybrid territories (chapter 5) of different forms. Yet they also clearly expressed that they belonged to and longed for Syria and Palestine. Even after years and decades of homemaking in Jordan, Syrian and Palestinian refugees have maintained strong senses of belonging to Syria and Palestine, the state-territories from which they, or their families, were displaced and from where they are now excluded. For some of these refugees, their senses of belonging to their "homeland" are more intense than they were when they lived there.

The territories of Syria and Palestine that they imagined certainly fit within the mold of conventional state-territories as being discrete and bounded. This was particularly evident for Palestinians, whose mappings of Historic Palestine reify the state-territorial nexus. Yet, these maps of Palestine concomitantly challenge the modern geopolitical ordering of the world from which Palestine is excluded. Syrians and Palestinians also imagined the state-territories from where they were displaced as ambiguous territories in which precise borders don't matter much. Likewise, symbols, emotions, and embodiments were part of how Syria and Palestine were imagined. These abstractions signify territories that are elsewhere, but a clearly defined territory is not required for these abstractions to invoke strong senses of longing and belonging.

Ambiguous and abstract imaginings of Syria and Palestine are not a rejection of the ordering of the world into discrete territorial states. Instead, they indicate the varied ways that the state-territorial nexus remains strong and meaningful, albeit without directly relying on imaginings of discrete, bounded territories. Relatedly, the fact that Syrians and Palestinians maintain such strong connections to territorial states that are British and French creations of the early twentieth century (Khalidi 2004, 67) is a strong testament to the continuing power and normalization of the state-territory nexus.

7

Refugee Camp Territories

One of the most ubiquitous and quintessential geographical images of forced displacement and refugees is that of squalid refugee camps. Photographs of overcrowded camps with desperate-looking refugees fill popular media reports on forced displacement and are featured on humanitarian web pages and fundraising materials. While refugee camps are typically poor and refugees feel desperate at times, neither camp life nor displacement trends fully fit this image. Instead, there are immense variations in the quality, form, and function of refugee camps and in the ways refugees experience life in a camp. In the three previous chapters, I focused on territorial imaginings and belongings at the rather large scales of pre- and anti-imperial territories, hybrid territories, and states. In this chapter, I narrow my scale and scope to refugee camps. Unlike pre- and anti-imperial territories, state-territories, and hybrid territories, which all exist in ways unrelated to forced displacement, refugee camps are spaces that are specific to displacement (Cole 2021).

There is an incredible variety of types and forms of refugee camps in Jordan. The two main Syrian camps in Jordan, Za'atari and Azraq (see figure 3.3), are fenced and highly securitized, remote territories that confine and segregate refugees by severely restricting their mobility. Refugees in these camps live in tents and caravans and have very limited opportunities to leave the camp. There are ten official UNRWA Palestinian refugee camps, and three more are run by the Jordanian government, scattered across populated areas of Jordan (see figure 3.2). These camps are markedly different from the Syrian ones. They are not fenced or securitized. Instead, Palestinian camps are open and human mobility in and out is unrestricted. Palestinian families live in modest concrete homes, and, while the camps are crowded and quite poor, quality of life in these camps is much higher than it is for Syrians in Za'atari or

Azraq. Regardless of these significant differences, there are also many similarities across the Syrian and Palestinian camps. All of Jordan's camps have clearly demarcated borders. Within the camps, Jordanian laws apply alongside additional rules and policies that are specific to the camps. Syrian and Palestinian camps are also humanitarian spaces, serviced by the UNHCR and UNRWA, respectively, in conjunction with the Jordanian government, the latter of which has ultimate authority over the camps.

While refugee camps in Jordan are numerous, only a small percentage of Jordan's refugees live within them: 18 percent of Palestinian and 16 percent of Syrian refugees in Jordan live in camps, both markedly lower figures than the global average of 40 percent.[1] Further, camps were not built for Iraqi refugees who came to Jordan in the hundreds of thousands after the US-led invasion of Iraq in 2003 (Bidinger et al. 2014, 71). Instead, Iraqi refugees (officially labeled "guests") have settled in Jordan's cities and towns. Thus, the large majority of refugees in Jordan reside outside of the camps.[2]

Camps are, in some regards, their own discrete territories with their own clearly demarcated borders and definitive rules on who is included and who is excluded. The camps all have definitive borders and some have walls, fences, and gates that clearly demarcate the camp and can segregate refugees. Yet, camps are also connected to countless adjacent and noncontiguous towns, cities, and communities. Even in the Syrian camps with securitized borders and strict monitoring of who comes and goes, the borders are permeable to human mobility, which creates a seeping effect outward from the camps' hardened borders (Sanyal 2014; Agier 2002; Darling 2017; Martin 2015; Abourahme 2015). This seeping effect is particularly notable for the open Palestinian camps, where refugees can come and go as they please as they have done for many decades. Thus, while refugee camps in Jordan are discrete and bounded territories, they are concomitantly seeping or "dasymetric," which is a cartographic term that refers to blurring and folding of spaces and territories. And finally, camps also exist in relation to the state-territories of Jordan, Syria, and Palestine. The camps do not form a bonded nexus with these states, but the camps have nevertheless evolved in relation to state-territories from which the refugees were displaced and the state-territory in which they were built. Refugee camps in Jordan, as I explain in this chapter, exist in relation to these three different forms of territory. They are their own discrete entities that mirror conventional territories; they are territories that exist relationally to Jordan, Syria, and Palestine; and they are networked, dasymetric territories that blur and seep into other communities. In the pages that follow, I first provide an introduction to the types of camps that have evolved globally since World War II, which is important context for understanding Jordan's campscape (Martin 2015). Then, stemming from interviews with Palestinian and Syrian refugees who live both in camps and in towns and cities in Jordan, I focus on three official camps and how they have blended different forms of territories together. I examine the Gaza and al-Husn Palestinian camps as their own discrete territories and then shift to demonstrate the ways that the camps exist in relation to different communities across the official borders, as well as to Palestine and to Jordan. I then discuss the Za'atari Syrian refugee camp, first as a territory in and of itself, then as a networked community, and finally as Syria and as Jordan. This chapter focuses on Jordan's complex campscape and the many ways the camps blur different territories, but more broadly I also argue that the existence of these camps demonstrates the failures of the state-territory-centric IRR and the durable solutions.

Varied Forms of Refugee Camps, Globally

The precise origin of refugee camps is unclear, but it is widely acknowledged that refugee camps grew in number during WWII, when tens of millions of people were displaced across Europe (McConnachie 2016; Arendt 1951). In 1950, the UNHCR was established to regulate and manage the refugees in Europe who had been displaced as a result of the war (see chapter 3). After WWII, with the rapid growth of newly independent territorial states, and conflicts erupting in some decolonizing states across Asia and Africa, mass displacements grew, and encampment became a common way to manage these new forced displacements (Chatty 2017, 180). Encampment is not one of the durable solutions, but since the mid-twentieth century, the UNHCR and the IRR more broadly have facilitated and supported mass encampments as a primary way to manage displacements across the globe (McConnachie 2016, 397).

The UNHCR (2013a, 12) defines a refugee camp as "any purpose-built, planned and managed location or spontaneous settlement where refugees are accommodated and receive assistance and services from government and humanitarian agencies." Theoretically and ideally, camps are temporary, built to provide short-term shelter and services during emergency situations. However, displacement often continues long after the "emergency" situation has ended, and this has resulted in camps becoming semipermanent, if not permanent, residences for millions of protracted refugees.

Official camps, which are ones that receive governmental and nongovernmental oversight and services, operate inside territorial states, but the laws that apply within the state do not necessarily apply within the camp. Instead, different rules and regulations govern refugees and refugee camps. In this way, camps can be the spatial materialization of what Agamben (2005) refers to as the "state of exception," or the condition in which the normal laws of the state do not apply.[3] With their own rules, refugee camps—like internment camps and prisons—are a form of biopolitical and territorial control. They are biopolitical in their practices of registering, documenting, ordering, monitoring, and controlling refugees. Concomitantly, refugee camps, particularly ones in rural areas or ones that are highly securitized, work as a type of territorial control by demarcating, fencing, and policing camp borders, as well as immobilizing and segregating refugees from the rest of the state-territory and its society. So while camps are often humanitarian spaces, they are also spaces of control and segregation. These dual roles—humanitarian assistance versus control and containment—seem like a contradiction, but in actuality, they work in tandem to create what Fassin (2012, 135) has referred to as spaces of "compassionate repression."

The UNHCR—along with many other organizations and governments—has long acknowledged the limitations and repressiveness of camps. The UNHCR admits that "the defining characteristic of a camp . . . is some degree of limitation on the rights and freedoms of refugees, such as their ability to move freely, choose where to live, work or open a business, cultivate land or access protection and services" (2013, 12). In 2009, the UNHCR pointedly encouraged settlement in cities and towns instead of camps (Ehrkamp 2017a, 817; Darling 2017). The organization's 2013 report on "alternatives to camps" states that its policy is "to avoid the establishment of refugee camps, wherever possible, while pursuing alternatives to camps that ensure refugees are protected and assisted effectively and enabled to achieve solutions" (UNHCR 2013, 6). Though the trend toward settlement in towns and cities is increasing in some states, other refugee-hosting states—including Jordan—have not followed this trend and instead have built new camps and maintained existing ones (Bernstein and Okello 2007; Fábos and Kibreab 2007). Indeed, camps remain a common way of managing and controlling refugees in many states.

Unofficial, informal, and makeshift refugee camps—those that are created without governmental oversight or regulated humanitarian services—are also quite common in some states. Martin, Minca, and Katz (2020) recently documented the growth of unofficial camps in Europe, like those along the "Serbian route" from Turkey to Europe and the so-called Jungle of Calais in France. Likewise, Sanyal (2017) has revealed the prevalence of unofficial camps in Lebanon, where the lack of state and humanitarian aid has forced refugees in these camps to become "self-reliant." Some unofficial camps will evolve into official camps with UNHCR services and state supervision, while others will be dissolved, and some persist as makeshift camps for years.

Palestinian Camps in Jordan

Palestinian refugee camps in Jordan (as well as in Lebanon, Syria, and the Occupied Territories) do not fit well with conventional imaginings of refugee camps (Ramadan 2012). These camps are not full of temporary, white tents, nor of desperate refugees. Instead, the camps have existed for as long as seventy years and look and feel like overcrowded small cities. Palestinian camps in Jordan were initially full of tents and other temporary structures, but they soon evolved. Most buildings and people's homes are made of concrete, while some "temporary" building materials (such as corrugated metal plates) still linger decades later. Far from temporary places of shelter, though, Palestinian camps have become what Alnsour and Meaton refer to as "permanent low-income housing settlements" (2014, 65). While there can be some speckling of wealth in Palestinian refugee camps in Jordan, overall, living conditions are substandard and overwhelmingly poor. Most of the buildings and homes in the Palestinian camps have just a few windows, are poorly ventilated, and are insufficiently insulated. Therefore, homes can be very hot in the summer and cold in the winter, as well as noisy and dark (Tiltnes and Zhang 2013, 16–17, 66). Over many decades, Palestinian camps in Jordan have had modest improvements, but they remain underdeveloped and need infrastructural improvements.

The lack of development in the camps is largely due to insufficient funding to improve them, but it is also partly intentional and political. Many Palestinians fear that improving living conditions in the camps or building "permanent" homes and communities could undermine their "right of return" to Palestine. The existence of the thirteen Palestinian camps and their substandard living conditions are materialized, daily reminders of Palestinians' protracted forced displacement, their daily struggles, and the lack of resolution with Israel.

The Palestinian camps, while poor, overcrowded, and politicized, are also spaces of home, family, friends, and community.[4] They are where tea, coffee, and meals are shared by generations of Palestinian refugees. They are where people laugh and watch television and kids go to school and play in the streets. As Ramadan (2012, 67) explains in his work on Palestinian camps in Lebanon, camps are "tremendously important and meaningful places for several generations of Palestinians who have known no other place to call home." Likewise in Jordan, the camps were built as temporary spaces of refuge, but they have evolved into homes and communities where life has endured for generations.

Of the ten UNRWA Palestinian camps in Jordan, four were established to accommodate the 1948 refugees and six were built for refugees of the 1967 war. UNRWA provides basic humanitarian services in the camps, including overseeing shelter, food, medical care, education, and economic and infrastructural development programs (Hanafi 2010; Robson 2017). Founded in December 1949, with the UN General Assembly Resolution 302, UNRWA predates the UNHCR and the 1951 convention. UNRWA's mandate is to provide aid, and not to engage in political activity or decisions of permanent settlement, including the durable solutions.[5] UNRWA works in tandem with the Jordanian state's Department of Palestinian Affairs (DPA) to administer the camps and provide services (Oesch 2016). There are three additional camps, Madaba, Sakhna, and al-Hassan, run fully by the Jordanian government. None of the thirteen camps are in isolated places, nor are they securitized with walls and gates. Instead, all thirteen of the Palestinian camps are open; meaning that human mobility and daily movement in and out are unfettered. While a few camps have walls or gates, those physical structures don't hinder mobility.

Palestinian refugee camps are, in some respects, their own distinct territories. They are all clearly demarcated on maps and most of them are also distinguishable in the landscape. Each individual camp also has its own identity, and its residents generally feel a sense of inclusion and belonging to their camp. However, each camp is concomitantly connected to adjacent and noncontiguous communities. Further, the camps exist relationally to Jordan and to Palestine. In the pages that follow, I frame Palestinian camps as four different forms of territory: as their own discrete, conventional territories; as dasymetric territories linked to many communities; as territories related to Jordan; and as territories related to Palestine. I discuss Palestinian camps in Jordan in general but provide a focused discussion of two particular camps: Gaza (Jerash) and al-Husn. Ultimately, I highlight both the extraordinary complexity of Palestinian camps and the myriad forms of territory that exist in relation to them.

CAMPS AS DISCRETE TERRITORIES

Palestinian refugee camps in Jordan have, in some ways, evolved into discrete territories. While they do not conform to all aspects of a conventional state-territory—most notably in that they lack independence—there are many other ways that the camps seem like territories of their own.

Jordan's thirteen Palestinian refugee camps all have official borders that cartographically demarcate the territory of the camp (figure 3.2). While the boundaries on the map are not always clear on the ground, most of the camps are distinguishable in the landscape due to the density, style, and layout of their buildings and roads.

Most residents of the camps maintain a sense of belonging to their particular camp. Many were born in the camps and have lived their whole lives there, and many have raised their families in the camps as well. Gabiam and Fiddian-Qasmiyeh (2017, 10) use the term "home-camps" to recognize the intimate connections and attachments that Palestinian refugees have with their camps. There is typically a strong sense of community within the camps that helps to facilitate such connections. A Palestinian man in his forties explained quite simply that "everyone knows everyone in the camps." A woman in her forties who had lived in Wihdat (New Amman Camp) for sixteen years but had moved out when she was married spoke nostalgically of her life in Wihdat. She had lived in a one-hundred-square-meter house with twenty-five family members. Regardless of the crowdedness of her home, she missed the camp and particularly its close-knit community. A woman in her fifties who had lived most of her life in Wihdat but had been living in Amman for nine years also reflected longingly on her life in the camp. She explained, "Life was nice in the camp. People there were all very connected and everyone knew everyone. It was just kind of different. It felt like its own small city, with its own shops, meat shops, and hospitals."

The Gaza camp, officially known as the Jerash camp, was established in 1968 as a result of the 1967 Six-Day War. The 1,500 tents that were put up in 1968 were replaced with prefabricated shelters not long after the camp's establishment.[6] As the decades passed, residents continued to modify and modestly improve the camp. Most buildings and homes are now made of concrete, but there are still houses made of old corrugated metal and some have tin roofs. Asbestos, a toxic powder that can cause respiratory problems and cancer, is not uncommon in homes in the camp (Marshood 2010, 69). There are approximately 560 homes in the Gaza camp that still have asbestos, and the rates of liver disease and cancer are very high there.[7] Some families have up to twenty-one people in them, but the average is closer to seven or eight. The camp is densely packed, the streets are narrow, and there are no open, green spaces. Until recently, the camp had open sewers from showers and sinks (not toilets) in the middle of some streets (Marshood 2010, 72; Tiltnes 2013, 27).

The Gaza camp has developed a strong and unique identity. The majority of the residents came from the Gaza Strip in 1967,[8] hence the common practice of referring to it as the "Gaza camp" as opposed to its official name of Jerash. All Palestinian camps in Jordan accommodate some Palestinians who are not Jordanian citizens; however, 94 percent of the 29,000 registered Palestinian refugees in the Gaza camp do not have Jordanian citizenship (Tiltnes and Zhang 2013).[9] As discussed in chapter 3, they are *nazeheen*, or displaced people who do not have citizenship or permanent residency in Jordan. These Palestinians are subject to different laws and regulations, they have different IDs and travel documents, and different educational and employment opportunities than Palestinians with Jordanian citizenship. The Gaza camp is the poorest among the ten UNRWA Palestinian camps in Jordan, with 52.7 percent of its refugees having an income below the national poverty line (Tiltnes and Zhang 2013).

Regardless of the challenges that Palestinians experience in Gaza, or perhaps because of the challenges, there is a strong sense of community and belonging to the camp. The leader of the camp, who is a refugee from the Gaza Strip, told me that "Gaza is unique; it is unlike any other camp." He explained that there was a strong sense of belonging to the camp, where people generally lived with their large, extended families. Most everyone knew most everyone else, he continued, and there were "very few outsiders" in the camp.

The Gaza camp is about a quarter square mile. It was built on private land that UNRWA rents (camp residents cannot own the property they live on). Like all the other official Palestinian camps in Jordan, Gaza has clearly defined borders. A framed map that hangs on the wall of the camp's main administrative office demarcates the spatial extent of the camp. The Gaza camp is also identifiable in the landscape. Its densely packed buildings, maze of streets, and lack of green spaces make this camp distinguishable from surrounding landscapes, which are a mix of farmland, residential homes, and some small businesses.

The al-Husn camp, which is often referred to as the Martyr Azmi al Mufti camp, was also established to provide shelter and care for Palestinians forcibly displaced during the 1967 war. Al-Husn is a little under a third of a square mile, it has clearly defined borders, and a framed map of the camp hangs on the wall in the main administrative office just as one does in the Gaza camp. I asked a camp leader, a Palestinian male in his fifties who worked with the DPA, whether there were any physical walls around the camp. He responded, laughing, "There are no walls—we are not Israel." Another male who was part of our conversation added, "We are not Za'atari either." The leader then explained that the outer streets of the camp are the physical markers of the camp's border. In other words, the on-the-ground physical border is an open road. There is an old gate that marks the main entrance to the camp, but it is only decorative and symbolic. Al-Husn is clearly distinguishable in the broader landscape, with its border road defining its edges, and with its

FIGURE 7.1. The landscape of al-Husn. Like many Palestinian camps in Jordan, al-Husn is clearly distinguishable in the landscape.

distinct style and density of buildings (figure 7.1). Much of the neighboring land is farmland, which provides a rather stark contrast to the crowded camp.

The camp's 15,000 original inhabitants were accommodated in tents, but in 1969, UNRWA built thousands of prefabricated shelters. Over many years, concrete buildings and homes have replaced the tents and shelters, but some sheet metal homes remain. Like other camps, al-Husn is largely poor and underdeveloped. Power goes out sporadically. Lack of trash removal is a common issue and many streets are littered. Medical care is available, but there are often long waits to see doctors. The leader of the camp estimated that each doctor in the camp sees seventy patients a day. There is a playground, a sports field, mosques, a cemetery, schools, and many stores and shops. There are now 25,000 registered Palestinians living in al-Husn, but unofficial population numbers range between 32,000 and 33,000 people. Among refugees in al-Husn, 23 percent have an income below the national poverty line (Tiltnes and Zhang 2013).

Most of the residents of al-Husn are refugees from the 1967 war who fled from the West Bank. At that time, Jordan controlled the West Bank and Palestinians living in the West Bank were Jordanian citizens. So, while these residents were forced to leave the West Bank as refugees, they were moving from one Jordanian territory to another. That status distinguishes these refugees from the 1948 refugees who were granted citizenship after their arrival in Jordan, and from the 1967 refugees who came from the Gaza Strip and were never granted citizenship.

As in other camps, most residents of al-Husn feel included within the camp and by fellow residents, many of whom are distant relatives. There is the sense in the camp, as one male in his fifties told me, that "everyone is close, everyone knows each other." An older male similarly told me, "We know most people and visit each other every day." The Palestinians I met in the camp stressed that there was immense impoverishment and a great need to develop the camps, but they also felt a strong sense of belonging to the camp and stressed that the people within were very close-knit. I heard from several residents in al-Husn that the only way they would ever leave the camp would be to move back to Palestine. The close-knit feel of the camp, its border road, and its distinguishable shape and extent all converge to create al-Husn as its own unique territory where so many Palestinians feel a strong sense of belonging.

CAMPS AS DASYMETRIC TERRITORIES

While their own discrete territories in many respects, Palestinian camps in Jordan are well connected "socially, economically, and urbanely" to their surroundings (Hanafi 2010, 20).[10] In protracted refugee situations, it is quite common for distinctions between camps, cities, and communities to blur (Woroniecka-Krzyzanowska 2017). Each Palestinian camp in Jordan is, to different degrees, connected and integrated to adjacent communities, as well as to communities that are noncontiguous and networked. Thus, while camps have defined borders and generally have a clear sense of who is included and who is excluded, camps also exist relationally to other forms of territories and communities.

I borrow the idea and term "dasymetric" from cartography as a metaphor to understand the relational and networked character of camp territories. Dasymetric mapping uses small spatial units of data (like tax lot level), fine spatial resolution, and/or point data in order to represent highly detailed geographic patterns at a micro level (Modeliranje et al. 2013). This type of mapping reveals the complexity and heterogeneity of sociospatial life by showing points of concentration and dispersion of a phenomenon in countless directions. Unlike the more common and similar technique of choropleth mapping, dasymetric mapping is not confined to formal administrative boundaries.[11] As camps have countless interconnections and relations to both adjacent and noncontiguous communities, I consider them to have dasymetric patterns.

Of the more than two million Palestinian refugees in Jordan, about 18 percent, or about 370,000 people, live in camps.[12] The camp residents are not forcibly segregated or confined to the boundaries of the camps. Refugees leave daily to socialize, shop, work, play, and study. Humanitarian workers and goods move in and out of the camps regularly as well. Camp residents also have the freedom to move to new homes outside of the camps and to move between different camps. Likewise, the 82 percent of Palestinian refugees in Jordan who live in towns and cities are not isolated from the camps or their residents. Non-camp refugees generally have family and friends in the camps and visit the camps to socialize, often on Fridays. Some non-camp residents go to the camps to shop for particular products or to access UNRWA health care centers. Many Palestinians outside of the camps once lived in a camp and on occasion, some refugees move from towns and cities into camps.[13] I interviewed an older man in al-Husn who lived in his house in Amman during the work week but kept his small house in the camp to live in on the weekends in order to socialize with friends and family. As discussed above, the al-Husn and Gaza camps are distinguishable in the landscape, but they are simultaneously part of neighboring communities. The Gaza camp is located about 50 miles north of Amman and is easily accessible by major roads. The camp is in the outskirts of the town of Jerash, about 3 miles from the University of Jerash and 3 miles from the famous Roman ruins of the same name. The al-Husn camp is located about 80 miles north of Amman; it is part of the town of al-Husn and less than 2 miles from the town center. Both these camps are at once separated from and connected to these major towns and countless other communities.

This blurring and seeping of the camps is not unique to al-Husn or Gaza. Two Palestinian women in their fifties, residents of the Marka/Hitten camp, laughed with each other, realizing that they could not figure out where the camp began and ended. They explained that there had once been a fence around the camp but that it had come down piece by piece as the camp expanded and new buildings were constructed. There is still an "iron gate" of Marka/Hitten, but it is just an area where streets converge. Indeed, it is difficult to find a beginning or an end to Marka/Hitten, whether by walking through the camp or looking at it from satellite imagery. A leader of the Marka/Hitten camp, a male in his forties, referred to the camp and surrounding city as "integrated and united." He explained that those in the city help to take care of the camp and that Palestinians outside the camp all "have the same troubles and the same dreams" as those in the camps. The blurring and diffuse interconnections between camps and other communities has not, however, extinguished the individuality of the camp or its official borders; instead, they coexist.

CAMPS AS PALESTINE

Since the first Palestinian refugee camps were established in Jordan more than seven decades ago, these camps have existed in relation to Palestine and have evolved both materially and symbolically as Palestinian territories. They are territories where both Palestine and a Palestinian identity are reconfirmed and rebuilt, and they are symbolic spaces that work as powerful reminders of Israel's occupation of Palestine and their right to return. After I arrived in a family's modest home in the al-Husn camp, the Palestinian man who lived there greeted me by saying "Welcome to Jordan," and then he winked at me and said, "Welcome to Palestine."

Most Palestinian families who fled in 1948 and 1967 were able to choose where they settled in Jordan, and many chose to settle with other families from the same village back in Palestine (Marshood 2010, 68). For example, several families in the Gaza camp are from Jabaliya in the northern part of the Gaza Strip; and many of those in al-Husn trace their roots back to Nablus in the West Bank. Their shared pasts and origins as Palestinians from specific villages has affected the built, material landscape in the camps. Refugees have designed parts of camps to replicate and commemorate the villages from where they were displaced (Dorai 2002). It is quite common that refugees use unofficial place names from back in Palestine to refer to the camps, as is the case with the Gaza camp, which is officially called Jerash. Further, refugees have named streets and areas within the camps after their home villages in Palestine. The settlement patterns and the place names within each of the camps certainly make each camp a bit unique and distinct, but they also concurrently recreate Palestine in Jordan.

Before 1970, some Palestinian camps in Jordan functioned like their own quasi state-territories. The PLO, under Yasser Arafat, controlled the camps and used them as headquarters for the PLO's political work. The Jordanian government was concerned with the strength of the PLO within Jordan and felt threatened by the quasi independence and subversive activity of the PLO (Farah 2008, 86). Tensions between the PLO and the Jordanian government erupted into a "civil war" from September 1970 to July 1971 (see chapter 3). During the war, known as Black September, the Jordanian government initiated martial law and the Jordanian armed forces targeted Palestinian camps.[14] Jordan's forces prevailed, ultimately exiling the PLO from Jordan and gaining direct control and oversight of the camps, which it maintains today.

The once palpable tensions between the Palestinian camps and the Jordanian government have subsided over the past several decades. Palestinian camps do not function as their own quasi-state-territories any longer; however, the camps remain important sites in the reproduction of Palestinian politics and identity (Feldman 2015, 247; Pappe 1994, 67; Achilli 2014). The camps are quintessential spaces for keeping the memory of Palestine alive and staying committed to the right of return. A man in the al-Husn camp explained, "Being here in the camp means concentrating on Palestine. Despite citizenship, despite having good lives and homes, in the camp we are always thinking of Palestine." Other Palestinians expressed that the camps facilitated more than just *thinking* about Palestine. Instead, the camps *felt* like Palestine. A young woman who had been born and raised in al-Husn explained that when she was in the camp she felt like a Palestinian living in Palestine. However, when outside the camp, she felt like an outsider living in exile. A young man who had been born and raised in the Gaza camp, and who had never been across the Jordan River to Palestine, imagined the camp to "feel just like Palestine."

Symbols of Palestine are common in the camps and work both to make the camp feel like Palestine and as daily reminders of the struggle to liberate Palestine. The Palestinian flag, map-logos of Palestine, and iconic images of Palestine, like the Dome of the Rock, Yasser Arafat, and artwork by famed Palestinian political cartoonist Naji al-Ali, can be found in homes and in public spaces in the camps. Three college-aged men explained to me that they wore their black-and-white kaffiyehs comfortably in the camps because, as one stated, "being in the camp is like being in Palestine." But elsewhere, including on their college campus in Irbid, they did not wear the kaffiyeh because, as a symbol of Palestinian identity, it might have drawn unwanted attention from Jordanians. Openly displaying a symbol of Palestinian identity is not uncommon outside of the camps, but it is more common in the camps, because the camps have evolved, in part, as a Palestinian territory.

CAMPS AS JORDAN

Palestinian refugee camps in Jordan are a part of Jordanian society, culture, economics, and politics. Jordan's independence in 1946 predates Palestinians' forced displacement by only two years, which means that Palestinian refugees and the thirteen camps that were established after the 1948 and 1967 wars are part and parcel of Jordan. Thus, while the camps are in some ways their own territory, are part of broader communities, and are recreations of Palestinian territory, there are also many ways in which Palestinian camps exist in relation to Jordan and feel and function like a Jordanian territory.

The application of law and governing practices are some of the most direct ways that demonstrate that the camps are a Jordanian territory. Palestinian camps do not fit the mold of Agamben's state of exception (Ramadan and Fregonese 2017; Feldman 2016), because normal Jordanian laws apply in Palestinian camps and camp residents are subject to Jordanian state law (Hanafi 2010; Alnsour and Meaton 2014). This includes Palestinians who do not have citizenship but are subject to Jordan's laws that apply to foreign nationals.

The Jordanian state, largely through the DPA, regulates and governs the camps (Hanafi 2010, 7–8; Oesch 2016). This includes policing and security measures. Water, sewer service, and electricity are provided by the Jordanian government in conjunction with UNRWA, which maintains a large, primarily humanitarian presence in the camps. Jordanian national education is taught in the UNRWA-run schools through tenth grade. Eleventh and twelfth grades, however, are both funded and run by the Jordanian state. The DPA appoints a camp representative to work with UNRWA who helps to manage the camp. While these leaders (often referred to as *mukhtars*) are Palestinian and have influence in the camps, they also function as "quasi-governmental employees" who largely answer to the DPA (Hanafi 2010, 13).

Jordanian national symbols are quite common in the camps and serve as banal reminders that the camps are part of Jordan (Culcasi 2016). Many administrative buildings in the camps fly Jordanian flags, and some buildings display the "Jordan First" and "We Are All Jordan" imagery. Framed photographs, posters, and paintings of the king and the royal family are likewise common in public spaces and in administrative offices. For example, in the Gaza camp, a "We Are All Jordan" sticker adorns a glass bookcase in the main administrative office, and the Jordanian flag flies near its entrance.

The DPA administrative office in al-Husn feels like an official Jordanian state building. There are photos of the late King Hussein, King Abdullah II, and the Crown Prince in this main office. Several Jordanian flags of different sizes are scattered around the room, while the Palestinian flag is conspicuously absent. These Jordanian symbols are notable in other areas of the camps as well. For example, at a girls' secondary school, a "We Are All Jordan" symbol and an image of the royal family adorn the entrance of the school, and the Jordanian flag flies on top of the building (figure 7.2).

More than just a symbol, the royal family has been regarded as a benefactor of the camps for many decades. At the start of King Abdullah II's reign, he gifted each camp furniture, refrigerators, and carpets. He recently gifted a theater to the Gaza camp and a cemetery to al-Husn. On his birthday and during Ramadan, the king often donates additional funds to the Palestinian camps. He also provides university scholarships to a limited number of camp

FIGURE 7.2. Girls secondary school in al-Husn. At the entrance of this school are three Jordanian symbols.

residents, including ones without citizenship. By supporting and aiding the camps, the kings have strengthened their relations with Palestinians in Jordan and have integrated the camps as a part of Jordan.

Syrian Camps in Jordan

There are five Syrian refugee camps in Jordan, all of which are in the north near the Syrian border. Zaʿatari and Azraq are the two largest camps. The three other camps are the Emirati-Jordanian camp, the King Abdullah Park, and Cyber City.[15] The UNHCR, along with many NGOs and INGOs, provides humanitarian services, like health care, education, and food, as well as some vocational training in the camps. The UNHCR actively seeks the durable solutions for Syrian refugees and serves as the "camp coordinator" for many of the camp's operations. The Jordanian state—through the Ministry of Interior, the Ministry of Planning, and the Syrian Refugee Affairs Department—oversees the camps at the broadest level and also provides its security measures (UNHCR 2021).

Unlike the Palestinian camps, Syrian refugee camps in Jordan do, in many ways, fit the Western imaginary of refugee camps. The Za'atari camp was built hastily in the summer of 2012 as tens of thousands of refugees crossed into Jordan (Francis 2015, 20). Za'atari was first filled with tens of thousands of tents, and over the years many of the tents have been replaced with prefabricated caravans. These two-hundred-square-foot temporary structures have walls, a floor, a ceiling, a small window, and a door that locks. They include a latrine and a tiny cooking area with a sink. The camp lacks a fresh water supply, and therefore water must be trucked in. Syrians I met both inside and outside the camp expressed grave concern about polluted water, long lines at the water tanks, and an overall shortage of clean water. The sand, rodents, tents, extreme seasonal temperature variations, lack of privacy, and shared bathrooms all make daily living difficult in the camp, if not miserable. Residents, particularly kids, frequently get sick because of these conditions. And while there are health centers in Za'atari, patients often have long waits. Frustrations about the conditions of the camps erupted into violence against some aid workers in the early months of the camp's opening. Some women have experienced gender-based violence in the camp, often as they walk to use the shared bathrooms or get water from the tanks. A woman I met who had left Za'atari after being there for a few months called life there a "nightmare." Another woman referred to the conditions as "inhumane." However, due to a lack of alternatives and to restrictions on their mobility, tens of thousands of Syrians have remade their lives in the squalid conditions of Za'atari. Many have been living there for years. And while their resilience and strength to persist is extraordinary, it is not surprising that Syrians expressed little to no sense of attachment or belonging to the camps. For the most part, they dreamed of returning home to Syria or at least of getting out of their camp and living in a modest apartment in one of Jordan's cities.

Azraq is the second major Syrian refugee camp in Jordan. It opened in 2014, making it two years newer than Za'atari. Both these camps are in isolated areas, but Azraq is more secluded than Za'atari. I did not do fieldwork in Azraq (or the three other small camps), but I learned from numerous aid workers and refugees, and through secondary research, that the conditions in Azraq were even harsher than those in Za'atari. This is despite the fact that Azraq camp designers had time and money for strategic planning that was not available when Za'atari was swiftly constructed two years earlier (Hoffmann 2017, 103). There are no tents in Azraq, and instead it is filled with immovable caravans that are anchored to the ground. Thus, refugees cannot alter the orderly layout, as they have in Za'atari. Azraq has no marketplaces or businesses, also unlike Za'atari, which has a thriving marketplace.

While built to provide refuge for Syrians, both Za'atari and Azraq are also designed to hinder the mobility of Syrians into Jordan, thus providing "security" for the Jordanian people and state (Gatter 2018, 23). As I discuss in the next section, Za'atari is highly securitized, with fences, gates, surveillance, and militarized guards.

In the next few pages, I describe Za'atari as blurring together four different forms of territory, mirroring those of Palestinian camps. I first discuss Za'atari as its own discrete territory. Second, I highlight the ways that the camp is connected to other communities, as a dasymetric territory. Only 16 percent of registered Syrian refugees in Jordan live in camps (UNHCR 2019), but many more have been there and left, whether legally or illegally. So while Za'atari isolated and securitized, it is also very connected and networked across Jordan. Third, I illustrate the ways that Za'atari has evolved into a small Syrian territory. And finally, I discuss how Za'atari exists in relation to Jordan, in the sense of its oversight, laws, and territory.

ZA'ATARI AS A DISCRETE TERRITORY

Just after Za'atari opened in 2012, it became the second-largest refugee camp in the world (by population) and one of the largest population centers in Jordan. In 2014, the camp hosted almost 200,000 refugees, its highest population since opening. As of 2021, there are fewer than 80,000 refugees in Za'atari, because many Syrians have fled the camp for better opportunities and improved quality of life elsewhere. In 2014, there were about 24,000 single-family caravans in the camp, thousands of tents, twenty-nine schools, three hospitals, nine health clinics, over a hundred mosques, some parks, recreation areas, community centers, and over a hundred NGOs and government entities (Ledwith 2014). In January 2021, there were twenty-six thousand caravans, which were slowly replacing tents (UNHCR 2021). There are also thousands of (unauthorized) refugee businesses, such as salons, bakeries, and electronic repair shops in Za'atari, most of which are clustered together and have formed a vibrant marketplace.[16]

Za'atari is about 1.5 square miles in area. Its extent and shape are clearly visible in the landscape and on satellite images, as it stands out starkly from the adjacent patchwork of scrub and desert. Its physical borders and the bordering practices of the Jordanian state have produced the camp as its own territory that segregates, confines, and sedentarizes refugees from the rest of Jordan (Gatter 2018; McConnachie 2016). Indeed, Syrian refugees do not have the freedom to move in and out of the camp, which is a violation of several different international human rights laws that guarantee the right to freedom of movement (Bidinger et al. 2014).[17]

There are several gates to the camp, each with very tight security. Rows of barbed wire fence line the perimeter of the camp, which are then encircled by a ring road. Beyond the fence and road, there are artificial sand berms that further hinder the mobility of refugees fleeing the camp, as sand hills are nearly impossible to run over discreetly or quickly. Armed forces and tanks are stationed near the sand berms, under canopy shelters that provide the soldiers protection from the sun and heat. Within the camp, barbed wire fences are also used to separate administrative areas from the rest of the camp. Around two hundred INGOs are working in Za'atari, and their offices are largely segmented off from refugees' housing, services, and business spaces (Tobin and Campbell 2016). Jordanian Public Security forces in the camp are highly militarized, with security officials driving SUVs reinforced with metal bars across the windows. The security officials monitor the inside of the camp and the outer perimeter. When I was in the Za'atari camp in the summer of 2014, fighter jets and helicopters flew in the skies over the camp on occasion, which served as a display of the strength of the Jordanian forces and the monitoring of the camp.

I was able to obtain a permit for myself and my three research assistants to enter the camp with only minor bureaucratic processing. However, because of the camp's security measures, exiting the camp proved to be more difficult. After a day of interviews and observations in the camp in the summer of 2014, my research team attempted to leave through one of the official gates, but we were detained for further scrutiny. The security officials suspected that one of my male assistants was actually a Syrian refugee whom we were smuggling out. It took ninety minutes to clear up the issue and prove his identity as a Jordanian national rather than a Syrian trying to flee the camp. This incident was, of course, frightening for my assistant and our entire team, and I decided to conduct more interviews in the towns and cities, as opposed to in Za'atari, in order to avoid any potential harm coming to my assistants. Yet this incident also clearly signaled that Jordanian security forces were pointedly on guard to stop Syrians from fleeing the camp into Jordan.

Other securitization and bordering practices have been implemented throughout Jordan in addition to encampment. While these practices are not necessarily part of the process of creating Za'atari as a discrete territory, they are nevertheless important to recognize as they are exemplary of how the Jordanian state has become increasingly restrictive to refugees. For example, all registered Syrian refugees have biometric iris scans as part of UNHCR's registration. While this practice helps increase the efficiency of registration, can help circumvent issues of identity fraud (UNHCR 2019), and allows for a cardless and cashless system of food assistance (UNHCR 2021), it is also a form of surveillance and biopolitical control over individuals both inside and outside of the camps. Likewise, the enforcement of *kefala*, as discussed in chapter 3, was implemented to restrict the mobility of Syrian refugees throughout Jordan and to monitor those refugees who were outside of the camps.

As noted above, Syrian camps exemplify Fassin's idea of "compassionate repression." In other words, the camps distribute humanitarian services and provide compassion and care, while concomitantly being territories of control, segregation, confinement, restriction, and repression. For example, the playground in the Za'atari camp shown in figure 7.3 is a quiet spot where kids may find a little reprieve and have some playtime, but this occurs under the surveillance of security forces in a tank, which is just beyond the playground fence, under a canopy. Expanding this contradiction of compassionate repression, McConnachie (2016, 406–8) argues that camps of this form "are sites that can perform humanitarian functions of protection, security, and service delivery while simultaneously serving political goals of containment and control. They are sites for the generation of human rights protection and for the abrogation or evasion of those protections." Emphasizing the latter, Za'atari (like Azraq) has been likened to "a prison facility" (Hoffmann 2017, 108) and "an outdoor prison" (Gatter 2018, 24).

In addition to hindering the movement of people within the camp, these repressive conditions also work to deter new refugees from coming to Jordan—much in the same way detention centers and family separation at the US/Mexico border were used by the Trump administration to deter migrants from even attempting entry into the US. The harshness of the camps is also an incentive for those Syrian refugees already in Jordan's towns and cities not to engage in any illegal activities, including working without a permit, because those who get caught violating Jordanian laws risk being sent to the camps (or being deported).

Za'atari is fully within Jordan's territorial borders, but it does not operate as an integrated part of Jordan. The physical borders of barbed wire and sand berms, and the repressive practices of segregation and containment in the camp, all make Za'atari seem like its own territorial enclave in the northern Jordanian desert. Unlike the Palestinian camps, Za'atari (like the other Syrian camps) is regulated by rules that are distinct from the rest of Jordan. Thus, Za'atari is in some ways an example of what Agamben (1998) refers to as a

FIGURE 7.3. Militarized playground at Za'atari. In this photo, a playground sponsored by UNICEF and Mercy Corp in the refugee camp is bounded by a fence and policed by a Jordanian tank in the background, exemplifying the practice of "compassionate repression."

"space of exception," or a space in which normal state laws are suspended during a time of emergency or crisis.

CAMPS AS DASYMETRIC TERRITORIES

Images of refugee camps may fill the Western image of displacement and refugees, and indeed there are discretely bounded Syrian camps in Jordan, but the effects of Syrian forced displacement in Jordan are dispersed, affecting countless aspects of life across Jordan. In other words, the camps are territorial in their segregation and containment of Syrians, but the effects and practices of establishing and running the Syrian refugee camps seep into surrounding communities and permeate throughout Jordan. In this section, I highlight some of the ways that the territory of Syrian camps intersects with life outside the camp. From its bounded, territorial location, the camps acts much like a hub, permeating outward in networked and dasymetric ways.

Many Syrian refugees in Jordan have lived in either Za'atari or Azraq at some point during their displacement. As of 2018, 461,701 Syrians had passed through Za'atari (UNHCR 2018c). The length of stays in the camps varies greatly. Some have been there for only a day or two, some for months, and some for years. Some Syrians have waited to obtain the legal paperwork that releases them from the camp, as part of the *kefala* system, but that paperwork can be difficult and expensive to obtain, and therefore many Syrians escape illegally, without the proper documents. Within the first year of Za'atari's opening, fence repair became a daily project in the camp because so many of its residents damaged the fences during their escapes (Ledwith 2014). In late 2013, the Ministry of Interior estimated that in the first eighteen months of its opening that 54,000 Syrians had escaped from Za'atari, often having been smuggled out in trucks carrying supplies and/or by bribing security guards (Ledwith 2014). A couple of Syrians whom I interviewed in Amman and Irbid mentioned that they had paid someone in the camp to help them escape. After two days in the camp, one woman I met had paid 200 Jordanian dinars (about US$280) to escape illegally. Reflecting on her two days in Za'atari, she described the urgency that she and her family had felt to flee the camp. She explained, "We needed to leave even though we weren't allowed to. We couldn't stay inside the camp and live in tents and suffer from the heat. So we decided to sneak out of the camp without the papers."

Some Syrians have had to return to the camps after living in Jordanian towns and cities. Sometimes they are forced to do so because they have engaged in illegal activity, like working without a permit, but other times they move back because they have no other options. Their family support systems or employment may change, leaving some unable to support themselves financially and with no choice but to return to the camp to receive basic humanitarian services.

Za'atari was built in the sparsely populated desert area of Za'atari, which is about 9 miles east of the town of Mafraq and 19 miles south of the Jaber border crossing. Unsurprisingly, the camp has had immense impacts on its immediate surroundings. Adjacent to its barbed wire fences and sand berms, the area immediately outside the camp has developed with new buildings and businesses. A bit further out, the village of Za'atari has grown as well, increasing its built-up area by 60 percent (al-Shoubaki 2017). Neighboring Mafraq has grown massively too. It is a busy town, with aid organizations operating there and humanitarian aid workers living there. Syrian refugees have also settled in Mafraq in large numbers. Between 2011, just before the forced migration of Syrians to Jordan, and late 2015, the population of Mafraq more than doubled, from 95,000 to 200,000 (Malkawi 2015). Mafraq has a substantial population of Syrian refugees, some of whom have lived and/or worked in the camp. For example, a wealthy Jordanian sheikh who owns apartments in Mafraq allowed widowed Syrian women to stay there, free of charge, while they got back on their feet. These women, who lived in about ten apartments, formed their own community, which they and the sheikh referred to as a "compound." One of the women in the compound whom I interviewed expressed her gratitude to the sheikh because she had been in the camp and was deeply aware that she fared much better in Mafraq than in the camp just miles away.

The city of Irbid (about 36 miles northwest of Za'atari) and the capital city of Amman (45 miles southwest of the camp) have both seen a huge increase in Syrian residents, as well as humanitarian aid workers, the NGO and INGO sectors, and the IRR more broadly. Many people in Irbid and Amman have either lived or worked in the camps and thus have networked interconnections with the camps. Likewise, across Jordan, water availability has decreased, rents have skyrocketed, trash removal has become more onerous, schools work on double shifts to accommodate all the children, and health care centers are overwhelmed. Thus the effects of the mass forced displacement of Syrians permeates far beyond the camp's fences.

Partly because of the poor conditions of the camps and the lack of better alternatives, some makeshift tented settlements have emerged on private and public lands. These unofficial or informal settlements often have UNHCR tents pitched on them, which were taken (illegally) from Za'atari or Azraq. For example, a makeshift camp about 12 miles south of Za'atari was settled by an extended family from Dera'a. The woman who was the leader of this camp had lived in Za'atari for three days before leaving. While meager, this camp gave these people the freedom of movement, something they did not have in Za'atari. About a hundred people, mostly women and children were living there in the summer of 2014 when I visited the settlement. They were sharing five large UNHCR tents. Some of the women worked for pay in neighboring farms, picking vegetables and olives during the harvest. They also had a modest garden and some goats. It is not entirely abnormal to see a single Syrian family living in UNHCR tents in yards and on private property (figure 7.4).

FIGURE 7.4. A Syrian family living in a UNHCR-issued tent in a Jordanian's yard. It is not uncommon for Syrians to smuggle tents out of the camps either to use themselves or to sell to others.

Unofficial, large camps and smaller single-family ones are not common in Jordan, at least compared to Lebanon, but they do appear on occasion as an alternative to the harsh conditions of camp life and the unfeasibility of affording an apartment.

Informal tent camps, the development of areas outside of Za'atari, the movement of so many Syrians in and out of the camps, and even the pervasive effects of hosting refugees and aid workers all demonstrate the ways that the refugee camps seep out of their official borders and have diverse effects across Jordan. Za'atari certainly has a clearly defined territory in which only certain people are included as refugees. Yet there is also an incredible flow of people in and out of the camp, as they seek better living conditions elsewhere. Za'atari seems to be fixed physically in one place, but its effects and reach are immensely broader.

ZA'ATARI AS SYRIA

Za'atari has evolved materially and symbolically into a little piece of Syria in Jordan. Syria and a Syrian identity are being remade in the camp through the close connections that people have with one another, as well as through symbols and symbolic spaces that are daily reminders of Syria.

Living with other Syrians in the confined space of Za'atari has led to the creation of many new and deep connection among Syrians. Syrians in the camp told me repeatedly that there was familiarity and comfort in being around other Syrians. Seventy-nine percent of Syrians in Za'atari are from the southern city of Dera'a (UNHCR 2018c). While they do not necessarily know each other from back home, their common place of origin has been a catalyst for building a sense of community in the camp. Many Syrians spoke of feeling supported by other people in the camp, explaining that Syrians went out of their way to help each other, often sharing what little clothes, food, or medicines they had. Due to their shared experiences as refugees, some Syrians built new friendships and felt as though they were closer to fellow Syrians in Jordan than they had been in Syria. Many Syrians I met referred to other Syrians in the camp as "family" or as "brothers and sisters." A sense of connection to other displaced people from the same place of origin is a common phenomenon among refugees and other transnational migrant communities. Building new relationships between people from the same "homeland" helps with transitioning, coping, and keeping the memories of home alive.

The building of Syrian refugee communities is happening in Jordanian towns and cities too. But outside the camps, Syrians regularly mix with and live among Jordanians, Palestinians, and Iraqis. Syrian camps and their residents are physically segregated from the rest of Jordanian society, and thus Syrians in Za'atari interact only with other Syrians (save security forces and humanitarian workers). While their confinement is repressive and abhorrent, it concomitantly buttresses meaningful connections between one another. I met one young woman in Za'atari who said that she preferred to stay in the camp over moving to a city because, as she explained, "in Za'atari all the people are Syrians. . . . In Jordan, outside the camp, I feel like I am a foreigner, but in Za'atari, I know all the people around me and I don't feel far away from home."

Besides a sense of connection with other Syrians that *can* evolve in the camp, there are other more symbolic ways that Syria is recreated in Za'atari. For example, Za'atari has an amazingly busy marketplace where there are approximately 3,000 (unauthorized) shops and businesses (UNHCR 2018). The main street of the market is known as Sham Élysées in reference to the famous Paris street Champs Élysées. *Sham* means "Syria" (as in *Bilad al-Sham*) in Arabic; thus, the colloquial naming of this street in Za'atari has the double meaning of being full of busy shops and also of being definitively Syrian.

A third pronounced way that the camps feel like a little piece of Syria in Jordan is through countless symbols of Syria and the Syrian opposition that are scattered throughout the camp. Examining 144 messages written on tents and caravans by camp residents in Za'atari, al Haj Eid (2019) found that the large majority of them expressed some form of patriotism, loyalty, or longing toward Syria. In addition to written messages, symbols such as flags, emblems, and logos are common in the camp, both in public spaces and within people's caravans and tents. For example, the flag of the Syrian opposition flies above some busy shops along Sham Élysées and can be seen on caravans. Co-opted for many other items too, like T-shirts and bracelets, symbols of Syria and of the opposition are commonplace across the camp. Such symbols work as daily reminders of Syria and help to produce the camp as a little piece of Syria in Jordan.

CAMPS AS JORDAN

Za'atari opened in the summer of 2012 and Azraq in 2014. Both were intended to be temporary camps, but as the Syrian war has dragged on and the durable solutions have yielded very few resolutions for Syrian refugees, tens of thousands of Syrians have been stuck in the camps. As a result, Za'atari has become a fixed territory in Jordan's campscape. Besides being physically within Jordan, and being under the control of the Jordanian state's oversight and policing, there is little about the camps that make them Jordan (hence the short length of this section). For example, unlike in the Palestinian camps, Jordanian national symbols are few and far between in Za'atari. Some Jordanian flags fly on administrative caravans and around gates, but their presence is quite understated throughout the camp. And the people in the camps certainly do not define themselves as Jordanian, nor did they express to me feeling of belonging to Jordan. Za'atari is a territory created entirely because of displacement. It is contained within Jordan and subject to Jordan laws, but it is territory of its own that seeps into surrounding and noncontiguous places, all while being a territory that exists in relation to Syria.

IN SUMMARY

Refugee camps are quite common in Jordan, but they differ immensely from one another, creating an amazingly complex and varied campscape. Palestinian camps are integrated in Jordan, and Palestinian refugees have unhindered mobility throughout Jordan. Syrian refugee camps, on the other hand, are securitized territories that segregate and confine refugees. These differences are significant, and, in some respects, the Palestinian and Syrian camps represent polar opposite ends of a spectrum of the types of camps. But these very different refugee camps also share many similarities.

Palestinian and Syrian camps exist in Jordan and the Jordanian government maintains control and oversight over all the camps. While the Syrian camps are much newer than the Palestinian ones, both Syrian and Palestinian camps were built as temporary spaces of refugee but have become more and more permanent. The quality of the camps certainly differs, but they are all quite poor and underdeveloped. While refugee camps often garner the attention of outside observers, only a small percentage of Syrian and Palestinian refugees actually live in the camps. And the most important similarity—as I explain more in the final paragraph below—is that the continued existence of the camps signals the failure of the durable solutions and the IRR.

Refugee camps in Jordan also exemplify the complexity of territory. They are their own discrete territories with their own histories and unique identities. Palestinian refugees generally have a sense of belonging to camps. While that is not the case with Syrian refugees, many Syrians have developed strong connections to their fellow camp residents. Refugee camps in Jordan are also dasymetric territories that are connected to (and sometimes blur into) adjacent and more distant communities through people's mobility and the creation of a wider campscape. Palestinian and Syrian camps are intimately connected to Palestine and Syria as well, creating camp territories that are material manifestations and symbolic reminders of their homelands on the other side of Jordan's western and northern borders. But all the camps are within Jordan's territory and are a part of the Jordanian state and society, albeit to different degrees. In these four different ways, camps reify conventional ideas of territory as bounded entities, but they also help reveal that there are other forms and scales of territory that are uncoupled from the state.

Whether a camp has a high degree of biopolitical and territorial control or is more open, camps are spaces of daily life for people who have lost their homes, have experienced trauma, and are living in displacement. Life is challenging and opportunities are limited in the camps, but refugees persist and life goes on. As Appadurai (1996a, 192–93) writes, life continues to unfold in camps as "marriages are contracted and celebrated, lives are begun and ended, social contracts made and honored, careers launched and broken, money made and spent, goods produced and exchanged."

Refugees live across all of Jordan. They are far from being contained within camps. Nevertheless, examining camps is still crucial to understanding forced displacement, as well as the broader IRR and "refugee crises." Further, the very existence of refugee camps and the increased territorial security and deterrence measures across the "Global North" and in Jordan over the past decade represent a global failure to guarantee human rights to people who have lost their homes and are seeking the most basic human need of safety (Singh 2020, 2). The practices of encampment—even as the UNHCR asserts the need for better options—and protracted use of camps blatantly demonstrate the failures of the durable solutions and the IRR more broadly (Martin, Minca, and Katz 2020, 753).

8

Conclusions

Crises and Failures

In the early twentieth century, European powers launched new programs and policies to address the mass displacements that happened in Europe after both world wars. The complex, global system that has evolved since then has helped provide safety for millions of displaced people, but it has concomitantly failed tens of millions more. The fact that 78 percent of refugees under the UNHCR's mandate remain in protracted displacement, that fewer than 1 percent of refugees are resettled annually in wealthy Global North states,[1] that 85 percent of refugees are in the Global South, and that refugee camps continue to be built and filled by displaced people each demonstrate the failures of the IRR (Hathaway 2018).[2]

Refugee "crises" are also an exemplar of the failures of the IRR. Europe's "refugee crisis," as discussed in the introductory chapter, did not happen because people were forcibly displaced due to war and persecution in Syria and elsewhere. Instead, war and persecution created the conditions which led to forced displacement. The "crisis" happened because several European states, as well as the broader IRR, neglected both their legal obligation to allow in asylum seekers and to provide reasonable and humane protection to asylum seekers and refugees. Since 2014, the term "refugee crisis" has become quite commonplace. It is used to refer to a "global crisis" as the sheer number of refugees worldwide continues to grow. It is also used to refer to crises at particular borders or locales, such as the US/Mexico border. Yet, for the most part, all refugee "crises" are a result of inadequate protections and/or humanitarian aid being granted to forcibly displaced people.

There are complex reasons that refugees are hindered from their legal right to seek safety and/or to receive humane protections. The lack of resources and wealth in some Global South states has created humanitarian aid "crises." Within many Global North states—notably Hungary, the US, and Australia—the discourse that refugees are potential terrorists, economic burdens, harbingers of cultural change, and demographic threats to white racial dominance (Gorman and Culcasi 2021) helps explain why some Global North states have gone to such extremes to securitize their borders and hinder asylum seekers from reaching them. Yet also underlying the deterrence practices and denials of entry, which have led to crises in the Global North, is the Western-invented ideal of the ordering of the world into territorial states and the related ideal that each person belongs to one state-territory.

Decades of globalization and the crumbling of physical borders with the end of the Cold War weakened the state-territory nexus to some degree and increased human mobility globally. However, the proliferation of territorial securitization projects, including the building of new walls and fences, over the past two decades show that the Western-centric state-territory nexus is strengthening. Our world of securitized territorial states makes movement and migration across international borders exceedingly difficult for most people who do not have powerful Western passports and the wealth to travel and move. It is generally easier for commodities like televisions or clothing to move across international borders than it is for people fleeing war or persecution. The state-territory nexus is also at the foundation of the massive international refugee regime. For example, a refugee can only exist in a world of discrete state-territories (Haddad 2003, 321), because, according to international refugee law, a displaced person must cross an international border to have status as a refugee. Similarly, the three durable solutions are each based on conventional, Western territorial ideals which view refugees as aberrant and therefore needing to be permanently reterritorialized into one state.

Even though most Global North states are signatories of the 1951 convention and/or the 1967 protocol, which are binding agreements, many signatory states quite freely disregard or violate the principles of these international refugee laws, and they do so without penalty. Some have domestic immigration laws and practices that are pointedly intended to hinder the movement of displaced people (and other migrants) into their state-territories, which is a direct contradiction of the 1951 and 1967 laws (Cole 2021; Dardiry 2017; Bose 2020; Jones 2021). Indeed, Global North countries spend four times more money deterring displaced people from accessing their territorial borders to seek safety than the UN has available to support all refugees living in the Global South (Hathaway 2018).[3] Some displaced people who try to reach Global North borders to seek asylum—like in the US and Australia—are treated like criminals and put in detention centers that function as prisons. Global North states also remain quite closed to the durable solution of refugee resettlement (Fassin 2012, 253). The US, for example, under the Trump administration reached new levels of disregard for the morals that are embedded in international refugee laws and human rights. Buttressed with anti-refugee and xenophobic discourses, security along the US borders with Mexico increased, asylum seekers were detained and refouled and forced to stay in Mexico, children were separated from their parents, COVID travel restrictions were implemented and retained, and the lowest ever annual refugee ceiling for resettlement was set in 2020. Further, just after taking office in January 2017, Trump issued an executive order that prohibited Syrian refugee resettlement in the US, a restriction that lasted until the Biden administration overturned Trump's executive order in January 2021. The territorial securitization practices at the US southern border and the bordering practices of banning Syrians from entering the US are testaments to the violent response that one Global North state has had to forced displacement. The situations and reactions in Europe in 2014 and 2015 differed, but in each case, the restrictive practices and securitization of territory led to crises. Meanwhile, Global South states—which are less likely to be signatories, generally have less wealth than the North, and are commonly perceived by Global North states as intolerant and authoritarian—are hosting 85 percent of the world's refugees.

The failures of the IRR are well acknowledged by scholars, humanitarian workers, the UNHCR, and, of course, refugees themselves. Nevertheless, the IRR and the durable solutions remain firmly in place. After the European "refugee crisis," there were concerted attempts to improve international practices. The New York Declaration of Refugees and Migrants in 2016 and the Global Compact on Refugees in 2018 were the results. While commendable, these are both nonbinding agreements which are largely reiterations of existing expectations, like the need for wealthy, Global North states to take on a greater proportion of the responsibility to host refugees and to provide support for refugees in becoming "self-reliant" through development programs. Neither of these agreements, unfortunately, has had widespread effects or facilitated substantial improvements (Chimni 2019; Hathaway 2018).

Learnings from Jordan

Jordan is not a signatory to the 1951 convention or the 1967 protocol, yet since its independence in 1946, Jordan has been quite open and welcoming to refugees. Today, Jordan has the second-largest refugee population per capita in the world. As it is a Global South state with high unemployment rates and a lack of natural resources, this is quite remarkable, and there is much to learn from Jordan.

Conditioned by anti-imperial geopolitics, regional geopolitics, and domestic politics, Jordan has responded differently to each wave of refugees. For example, 1948 Palestinian refugees were granted citizenship in Jordan; and that status has given Palestinian Jordanians political rights and the opportunity to thrive as business owners and entrepreneurs. Their contrast to 1967 Palestinian "guests," who do not have citizenship and who are much more likely to live in poverty, serves as a testament to the effectiveness of citizenship rights in facilitating people's chance to prosper. Waves of Iraqi and Syrian refugees, as well as smaller flows of other populations, have been treated largely like 1967 Palestinian refugees, as long-term guests who have been de facto integrated into Jordan.

Integration has not been an official durable solution in Jordan since 1998; nevertheless, de facto integration has been incredibly common. Jordan's widespread practice of allowing long-term stays for refugees into its towns and cities (as opposed to encampment or refoulement) has also allowed countless Palestinian, Iraqi, and Syrian refugees to maintain some sense of independence and dignity. Integration does not secure political rights like citizenship, but it does give a person their mobility, opportunities to settle and secure residency, job opportunities, and access to the public education and health care systems. While not a perfect solution, if there is even such a thing, de facto integration has offered many opportunities for refugees in Jordan to rebuild their lives.

While refugee practices in Jordan have been rather open and are commendable, since 2011, Jordan's policies and practices have shifted toward prioritizing securitization of their state-territory over the human rights of refugees. The once quite open borders and compassionate policies now coexist with repressive state practices, including the forced encampment of some Syrians and the closure of the Syrian border. Jordan once seemed like a "haven" (Chatelard 2010b) for refugees, but its more open practices have waned and now mix with ones that mirror the deterrence strategies of the Global North.

Jordan is a postcolonial state with imperially drawn borders that have no pre-imperial historical, cultural, or political meaning. Both the Jordanian government's policies and practices and the experiences of Palestinian and Syrian refugees in Jordan demonstrate that pre-imperial and anti-imperial territorial imaginings (chapter 4), the modern state-territorial form (chapter 6), and the mixing of the two (chapter 4) continue to affect government policies and refugees' multifaceted experiences with belonging, as well as their decisions about moving and staying. As postcolonial scholarship reminds us, it is crucial to consider the ways that imperialism lingers in the present. Likewise, as feminist geopolitics and decolonial studies assert, it is also necessary to examine and bring attention to histories and geographies that are too often sidelined as a result of the normalization of the modern world order. As a Palestinian woman in her sixties told me, "the Western world has tried to convince us that borders are necessary," but the woman rejected this idea. She had lived in Amman since she was a young girl, when her parents fled Palestine in 1967. She loved Jordan. She had been raised there and then had raised her own family in Jordan too. She referred to her life in Jordan as "wonderful." But she considered the border between Palestine and Jordan to be a "tragedy." She elaborated that it was a "Western power grab" that had countless negative effects on Palestinians, including curtailing her ability to visit her family on the other side of the Jordan River. She wished to return to the way the land had been before Western powers intervened, as an open and fluid territory.

While challenges to the imperial borders and the Western-based state-territorial form are significant, the state-territory nexus nevertheless remains strong and integral to Palestinians' and Syrians' territorial imaginings and senses of belonging (see chapter 6), as well as to Jordanian laws and practices (see chapter 3). Connections to and longings for their homeland, where they or their families were displaced from, reify the state-territorial norm as well as the specific borders that French and British imperialists created in the aftermath of the fall of the Ottoman Empire. That conventional state-territorial imaginings are so strong among refugees in Jordan clearly indicates the extraordinary power of state-territories.

Refugee camps are not an exception in Jordan but instead are quite commonplace. The Jordanian government has built camps for Palestinian and Syrian refugees, and these camps have evolved in tremendously different ways, with the former being open while the latter work as detention centers. Examining Jordan's refugee camps reveals that camps are extraordinarily complex and varied. However, only a small percentage of refugees in Jordan live in camps; thus, the complex campscape in Jordan also teaches us that it is crucial to study the many forms of refugee territories more generally, as opposed to only camps.

Moving Forward

Prioritizing global collaboration and human rights over state-territorial security is central to improving the IRR and stopping "refugee crises." While the preferences for collaboration and rights are already part of international refugee laws and part of many IRR practices, they are nevertheless commonly sidelined in the interest of securitizing state-territories. Below I suggest several ways to move forward with improving the IRR and, most importantly, the lives of displaced people. My suggestions include both broad discursive changes and smaller, piecemeal ones as well. Western powers created the IRR only about a hundred years ago. The state-territory nexus has not existed for much longer. There is nothing permanent about either. They both can change.

CHALLENGING THE STATE-TERRITORY NEXUS

To make substantial changes to the IRR, there needs to be broad recognition of the Western creation of the modern world order and its centrality to refugee policies, practices, and lives, as well as the associated ideal that people are sedentary citizens of only one state. This includes critiquing the imperial foundations of the majority of the world's territorial states and the challenges and injustices this ordering creates for displaced people attempting to move for safety. While I did not tease out the connections pointedly, it is important to note that neo-imperial, settler colonial, xenophobic, and racist discourses are integral to broader imperialist foundations that linger today.

As literature from critical geopolitics and postcolonial studies has asserted for decades, there is a need to examine the taken-for-granted assumption that the world is divided and ordered into state-territories, as this can help to show that there are other forms and scales of territory besides the Western state-territory nexus. While Western powers (including the IRR) have erased or subordinated many pre-imperial, anti-imperial, and nonstate forms of territory, territorial imaginings and senses of belonging based on nonstate-territories have always existed and continue to impact refugee policies and lives. As postcolonial scholar Dipesh Chakrabarty reminds us, there are "many ways of being-in-the-world" (2000, 255). A Syrian woman in her early thirties exemplified Chakrabarty's assertion, as she so simply stated, "I do not care about states or nationalities; I deal with humans." Recognizing that people maintain different and multiple connections besides those to state-territories can help facilitate meaningful changes to policy and practice that would allow displaced people more mobility and more opportunities to settle and resettle in a variety of places. Weakening the state-territory nexus can also help improve global collaboration by stressing the interconnectedness of territories and lessening the practices of prioritizing state-territorial borders over people's lives.

Moving away from the state-territorial policies may seem undesirable, or even impossible, to many people in the Global North, particularly as anti-immigration and anti-refugee discourses, politics, and practices have been mainstreamed. But the dominant Western-based state-territorial nexus has decidedly imperial origins that do not match the policies, practice, and experiences of countless postcolonial states and the refugees who have sought safety across them. Not only do many refugees imagine territorial entities and maintain senses of belonging that do not fit the modern state-centric ordering of the world, but there are also policies, practices, and theories already in place, as noted in chapter 2, that are less state-territory-centric. Often referred to as "flexible," "mobile," "regional," and "transnational" approaches, these differ to some extent, but each prioritizes securing human rights—like family reunification, residency, employment, and mobility—within and across borders (Long 2014; Ong 1999; Gabiam 2016). These approaches support refugees' rights to sustain their livelihoods, but they do so without requiring that the refugee be a member of one, singular territorial state.

CHALLENGING THE NEGATIVE DISCOURSE ON REFUGEES

It is essential that the negative discourse around asylum seekers and refugees (and other migrants) changes in the Global North. While refugees are certainly treated with compassion and kindness by many individuals, groups, and governments, much of the world has antipathy, if not hostility, toward displaced people (and migrants). In the Global North, refugees are all too often discursively constructed as objects of fear, as criminals, threats, and terrorists, and as harbingers of unwanted cultural change. They are often framed as burdens who bleed welfare systems, taking limited resources and jobs away from citizens. This negative discourse stems from xenophobia, fervent nationalism, and racist ideologies that prioritize protecting a state-territory and one's dominance within it over other human beings. But there are many reports and studies that have documented the strength, success, and independence of refugees who have settled after fleeing war and persecution (Mathema 2018; Anderson 2022). Thus, the negative discourse on refugees stems more from racist, exclusionary politics and values in the Global North than it does from the reality of refugees' impact on hosting societies.

In writing this book, I hope to have helped humanize refugees and counter the discourse that refugees are a threat or burden, because this discourse is all too common in the Global North and is used to justify state-territorial securitization and violence against displaced people. While I have not provided rich or thick details about refugees' experiences, nevertheless, the preceding chapters show that refugees in Jordan are not aberrant, exceptions to the norm, nor are they seen as threats or as dangerous to Jordanian society. Refugees are not merely victims or burdens either, but amazingly resilient people who have survived trauma and loss. Palestinian and Syrian refugees are people who struggle, cope, and adjust. They are people with different pasts, presents, and futures. Like refugees across the globe, they have the resilience, ingenuity, and agency to remake their lives in challenging circumstances (McConnachie 2016, 406–8). The Palestinian and Syrian refugees I interviewed are some of the most resilient, capable, generous, and welcoming people I've ever had the honor of meeting. I would be pleased to have any of them as my friend, neighbor, or colleague. Humanizing forcibly displaced people and challenging the discourse that they are threats or burdens can lead to more humane and ethical approaches to refugees in the Global North (and in the South), including providing access to borders to seek asylum and increasing refugee resettlement in wealthy states.

INCREASED MOBILITY

Within the current IRR, forcibly displaced people must be able to reach international borders and cross them in order to officially seek asylum and to gain status and protections under the UN Convention Relating to the Status of Refugees. Hindering displaced peoples' access to territorial borders where they could seek asylum, under the guise of protecting a state-territory, exacerbates the struggles and suffering that displaced people have already experienced as the result of war and persecution. Instating domestic laws and practices that allow forcibly displaced people to reach a state's borders and to formally seek asylum is one of the best ways to help improve the safety and well-being of displaced people. An increase in *planned* mobility, in the form the durable solution of third-country resettlement, could also help refugees to rebuild their lives. Even temporary resettlement through flexible, humanitarian, student, work, and family reunification visas or policies creates opportunities for refugees to manage their trauma and to improve their lives.

There also needs to be greater mobility *within* states for both refugees and asylum seekers. People need to be able to move within a state to find employment and education opportunities, to reunite with families and friends, and to maintain their senses of independence and dignity. Detention centers and refugee camps, which both severely hinder movement, should be abolished as a violation of human rights and an insult to personal dignity.

ROOT CAUSES?

Ending conflict, war, and persecution so that displacement does not occur in the first place is, of course, ideal. Often referred to as the "root causes" of mass displacement, the factors that lead to displacement certainly need to be addressed. But this is an epic challenge and one that has no easy or agreed-upon path. Solving root causes of displacement becomes even more difficult when other interconnected factors like endemic poverty, corruption, lack of political and human rights, environmental disasters, and climate change are given full recognition. Mass, forced displacements that are not directly linked to war and persecution do not meet the criteria for aid and protections as established by international refugee laws, yet they still cause mass displacement and human suffering. So while root causes matter and should be addressed, a broad and nuanced view of complex and interconnected causes of displacement is needed. Further, until we figure out a way to stop war, conflict, global climate change, and endemic poverty that can lead to forced displacement, fixing the IRR needs to be a priority.

Additionally, fixing "root causes" must include recognizing and rectifying the roles of Western imperialism, military intervention, and capitalist exploitation in both creating and exacerbating the conditions that have led to many mass displacements (Chimni 2019). For example, in order to slow the flow of asylum seekers trying to enter the US, US Vice President Harris traveled to Guatemala and Mexico in June 2021, announcing that the US would provide US$40 million in humanitarian aid to improve the "root causes" of displacement from those countries. Concurrent with her announcement of this aid, she also stated during a news conference in Mexico that potential asylum seekers should "not come" to the US and that they would be "turned back." Her statements ignored US domestic refugee laws that have existed since 1980 and the US's obligations to grant asylum as a signatory to the 1967 UN protocol. Moreover, her anti-immigration discourse also patently ignored the US's direct and complicit role in stoking conflicts in Central America in the 1980s, which created the conditions that have led to the endemic poverty, crime, corruption, and gang violence from which tens of thousands of people from Guatemala, El Salvador, and Honduras are attempting to flee. Perhaps recognizing the US's role in the protracted violence and poverty in Central America could lead to greater acceptance of these asylum seekers and to more sustained aid that could help alleviate the violence and destitution that has caused so much suffering and displacement.

There are countless other examples of how Western states—whether during the heyday of colonialism, the Cold War, or the "global war on terror"—are part of the "root causes" of mass displacements in the Global South. Shifting focus back to Jordan, Western powers have been directly and indirectly part of the root causes of mass displacements into Jordan. The US's military interventions in Iraq in 1991 and 2003 directly created mass displacements of Iraqis into Jordan. While the immense violence of ISIS in 2014 was an immediate cause of the mass displacements of Iraqis and Syrians into Jordan in that year, it is important to remember that ISIS grew out of the US's invasion of Iraq in 2003. Yemeni refugees in Jordan are also an indirect result of US foreign policy, which has turned a "blind eye" to Saudi Arabia's attacks on Yemen. Likewise, many Afghan refugees are in Jordan as a result of decades of protracted conflict in Afghanistan, including twenty years of American occupation. And while not the most immediate cause of conflict in Israel/Palestine, the US has been complicit in the protracted refugee situation of Palestinians. The US's unequivocal support of Israel has kept Palestinians from returning to Palestine. Under the Trump administration, support for Israel led to the US completely cutting off funding to UNRWA. The 2011 Syrian war was not directly connected to US geopolitics, but nevertheless, the US initially supported the uprising and trained opposition Syrian fighters, while doing little to negotiate or foment peace with Assad and his Russian supporters.

As noted in chapter 4, the British- and French-drawn borders after WWI are not the singular cause of Syrian, Iraqi, or Palestinian mass displacement. However, those imperial borders created the state-territories of Syria, Iraq, Jordan, and Israel/Palestine from the desires and purview of a few European powers and not of the people who lived there. These borders, which have existed for only one hundred years, are the same ones that millions of refugees have crossed since the mid-twentieth century in order to seek protections from conflict and persecution in their imperially constructed state-territories.

In Conclusion

Fixing root causes of conflict that leads to displacement is important. So too is stopping "refugee crises" from happening as wars and conflict continue. Creating a more collaborative and humane IRR that is less based on the state-territory nexus, that promotes a positive discourse on refugees, and that allows for greater refugee mobility could greatly improve lives and opportunities for displaced people. While working toward these immense goals, there are also some immediate, albeit piecemeal, changes that can be made to improve the lives of refugees. For example, providing more aid to the Global South states that are hosting the majority of the world's refugees is needed. Increased mental health and psychosocial support could also help refugees to cope and become independent. The right to work should be guaranteed, since working leads to self-reliance and dignity. COVID has, of course, caused major problems and has exacerbated preexisting social and economic hardships for many refugees. Widespread vaccination distribution is greatly needed as a first step to combating the virus among refugees.

Research on migration and displacement must continue to examine dominant power relations and underscore the inequities that are produced. I have drawn primarily from critical geopolitics, feminist geopolitics, postcolonial studies, and decolonial studies in order to examine the role of territory in the IRR, to help critique unequal power relations, and to help reveal subordinated geographies and histories (Collins 2022). Such approaches also need to be better integrated into refugee studies, into public discourse, into educational curriculum. Such lenses can offer new ideas and options for addressing the many refugee crises that exist. Bias, stereotypes, misinformation, and discrimination stem—in many cases—from a lack of experience, interaction, and education.

Finally, as I sit comfortably in my home in West Virginia, secure in my citizenship status and with my associated political and human rights (but no longer my reproductive rights), I write this conclusion with much consternation. The huge majority of Syrians and Palestinians I met in Jordan remain in protracted displacement. Many of them continue to struggle, but it is important to underscore that they are also full of strength and ingenuity. Further, many of them have powerful and informed ideas about how to improve the system that has so much influence on their lives. While I work earnestly to educate people (mostly in the Global North) about refugee lives and the IRR, refugee voices and leadership are greatly needed. Likewise, people living in the Global South, where the majority of refugees live, need to be central to such conversations that are dominated by people in the Global North. This book stems from interviews with refugees in Jordan, but like most Global North researchers, I am not a refugee. Hearing from refugees themselves is an essential next step to better learn how to improve the system (Chatty 2016). Many organizations highlight refugee "voices" and stories (including the UNHCR), but we need to give refugees control over their representations (Godin and Doná 2016), which requires us to be humble researchers (Cabot 2016), and to support promoting refugees into leadership and researcher positions (Gatrell et al. 2021). Who else could better know about the problems and potential solutions of forced displacement than refugees themselves?

Notes

Introduction

1. Hannah Arendt (1951), in *The Origins of Totalitarianism*, argued powerfully that refugees exist because states fail to protect people.

2. The term "open" can be used in various ways, but it often signifies a state waiving visa requirements for entry and residence permits for migrants to stay within the territory for a set time (Bidinger et al. 2014).

3. Paul Currion, "U.N. Refugee Summit's Numbers Do Not Add Up to Reality," Refugees Deeply, September 23, 2016, https://www.newsdeeply.com/refugees/community/2016/09/13/u-n-refugee-summits-numbers-do-not-add-up-to-reality.

4. There are many spaces of refuge (Cole 2021), like transit and registration centers, health clinics and schools, restaurants and other businesses, and INGO and NGO centers, settled communities, compounds, informal camps, and housing complexes. These spaces operate less like territories than, for example, a state or a camp, but they are still integral spaces of the IRR and of refugees' lives.

5. The political division of territory into states has cultivated a world, at least theoretically, in which people derive their legal place of belonging as a citizen of a state. While citizenship as a legal status theoretically guarantees a person's rights and inclusion in their state, citizenship often does not secure rights, nor does it guarantee inclusion or equality.

6. I conducted a total of sixty-nine interviews with Syrian refugees or with experts working with Syrian refugees. Fifty were with Syrian women (two of these were follow-up interviews in consecutive years), ten were with Syrian men, and nine were "expert" interviews with INGO workers and government officials (I consider an "expert" to be someone whose profession is working with refugees, and thus I asked them more reflective questions about policies than about their experiences and imaginings). I conducted a total of fifty-seven interviews with Palestinians or experts working with Palestinian refugees. Twenty-three were with Palestinian women; twenty-four were with Palestinian men, and ten were expert interviews.

7. Discussions sometimes included comparisons between Syrians and Palestinians. There was a common sense that Palestinians and Syrians understood one another as fellow refugees. However, there were also tensions, including Palestinians' occasionally feeling that Syrians were taking limited state resources from them (and Jordanians), and some Syrians felt that Palestinian (and Jordanian) landlords and employers were exploiting them. Experiences of male and

female refugees are notably different; and there is a growing body of literature on gendered differences of displacement (Horn and Parekh 2018).

8. Martin (2015) coined the term "campscape" in reference to Palestinian camps in Lebanon. I borrow her term but apply it more broadly to refer to camps and communities across Jordan.

Chapter One

1. "Modern" is a famously problematic term, but it generally refers to a period since the "Enlightenment" in Europe, when industrialization and the reduction of church power altered political, economic, and social relations.

2. For more examples of examining politics at the micro scale , see literature on memory, affect, and emotion (Tolia-Kelly 2004; Cain, Meares, and Read 2015) and feminist international relations (Sylvester 2012; Enloe 2000; Cockburn 2007).

3. While I do not discuss the gendering of territory, it is important to note that territory is often feminized as the "motherland," as giving life and nurturing. It is also masculinized as the "fatherland," as protecting and overseeing its people. See Ramaswamy's (2010) fascinating book *Goddess and the Nation* on the gendering of India for a detailed examination of the gendering of India.

4. Another reason I focus on the state-territory concept is to move away from using the term "nation-state," which I agree is an utter fiction. Iceland and Japan are often considered the only nation-states that exist, yet the term and the concept are uncritically accepted.

Chapter Two

1. There is nothing new or abnormal about human migration. Migration has been a constant and fundamental part of human history. Sedentary life (staying and living primarily in one place) has been the norm across the globe only quite recently. With the emergence of "modern" territory and the dominance of the state-territorial nexus in the past few hundred years, migration was qualified as a problematic issue to states. In other words, it is within this recent context of sedentary life that migrants and displaced people have been considered a problem more than the norm. As Mountz (2009, 174) writes, the norm of sedentary life within a state led to migrants and refugees being "categorized, politicized, qualified, quantified, studied and controlled."

2. The shift from the Nansen passports to the 1951 system marks an important moment in which refugees became a category separate from economic migrants, as the former focused on work, employment, and reestablishing livelihoods and the latter minimized labor and focused more on aid (Chatty 2017, 179; Long 2013).

3. The lack of protection from individual states has led to the need for the UNHCR and other NGOs to provide humanitarian assistance (Betts 2013; Lopez, Bhungalia, and Newhouse 2015).

4. During the Cold War, Western capitalist states viewed refugees as markers of Western freedom and liberty versus communist restrictions and were particularly willing to resettle refugees fleeing communist states. After the Cold War, Western states became more reluctant to receive refugees (Long 2013, 21, 106; Fassin 2012, 224).

5. See UNHCR, "Global Trends: Forced Displacement in 2018," https://www.unhcr.org/globaltrends2018/. That percentage does not include Palestinians under UNRWA, but it does include Palestinian in Egypt as they are not under UNRWA's oversight.

6. See also UNHCR, *Handbook on Voluntary Repatriation: International Protection*, https://www.unhcr.org/en-us/publications/legal/3bfe68d32/handbook-voluntary-repatriation-international-protection.html.

7. UNHCR, "Global Trends: Forced Displacement in 2018," https://www.unhcr.org/globaltrends2018/.

8. UNHCR, "Solutions," https://www.unhcr.org/en-us/solutions.html.

9. UNHCR, "Regional Resettlement Data," https://www.unhcr.org/en-au/regional-resettlement-data.html. In the 1990s, the UNHCR began to emphasize assisting and settling refugees in their region (as opposed to resettling them in distant states), thus containing refugees locally. Though there are benefits of staying near to the place of displacement, this approach also greatly diminishes the responsibility of wealthy Western states to resettle refugees in their own territories.

10. Nick Miroff, "Trump Cuts Refugee Cap to Lowest Level Ever, Depicts Them on Campaign Trail as a Threat and Burden," *Washington Post*, October 1, 2020, https://www.washingtonpost.com/immigration/trump-cuts-refugee-cap/2020/10/01/a5113b62-03ed-11eb-8879-7663b816bfa5_story.html; "Statement by President Joe Biden on Refugee Admissions," May 3, 2021, https://www.whitehouse.gov/briefing-room/statements-releases/2021/05/03/statement-by-president-joe-biden-on-refugee-admissions/.

Chapter Three

1. Displacement also happens because of development projects like the building dams or urban development. Climate and environmental displacements (e.g., flood, fires, drought) are increasingly prevalent.

2. Another major migration occurred when 200,000–300,000 Jordanian Palestinians returned to Jordan from Kuwait during the 1990 Gulf War. International relations between Jordan and Kuwait soured due to Jordan and the PLO's support of Iraq, leading to this expulsion (Massad 2001, 243).

3. For example, see the SWANA Alliance website at https://swanaalliance.com/.

4. Jordan is a constitutional, parliamentary monarchy with an elected lower house (Chamber of Deputies) and an upper house (Assembly of Senators) that is appointed by the King. The King appoints judges and the Prime Minister. He is also the head of the army and the head of the state. With so much power centralized within the monarch, Jordan is often characterized as a "partial" or "paper" democracy (Milton and Hinchcliffe 2009, 9). During the Arab revolts that began in 2011 that shook established ruling powers in many neighboring Arab states, there were only modest protests in Jordan. Some Jordanians certainly have grievances, but their issues are generally directed towards the government and the economy, rather than to the ruling family.

5. Kate Lyons, "Six Wealthiest Countries Host Less Than 9% of World's Refugees," *The Guardian*, July 17, 2016, https://www.theguardian.com/world/2016/jul/18/refugees-us-china-japan-germany-france-uk-host-9-per-cent.

6. The label "West Bank" was adopted in 1950, replacing several other terms like "Palestine," "Central Palestine," and the "Western Territories" (Massad 2001, 229).

7. The UNHCR began operations in Jordan in 1991 when Iraqis were fleeing the Gulf War, but not until 1997 did the organization open an office and form its relationship with Jordan (Hilal and Samy 2008).

8. Jordan's 1952 constitution (specifically, article 21) likewise prohibits the extradition or refoulement of "political" refugees.

9. In March 2014, three years after the start of the Syrian war, the 1998 MOU was modified slightly to give more time for registration and renewals, but it preserved the original principles.

10. Jordan hosts the largest overall number of Palestinian refugees, and the others follow respectively from highest to lowest. See UNRWA, "Where We Work," https://www.unrwa.org/where-we-work. Natural demographic growth has accounted for much of the increase in refugees since 1948.

11. For decades, electoral divisions across Jordan have favored rural Bedouin, East Bank Jordanians over the more urban-based Palestinian electorate. One of the several reasons that 1967 refugees remain without citizenship is because granting them citizenship would increase the Palestinian electorate and would likely affect elections in favor of Palestinians, thus shifting the power of the East Bank Jordanians toward Palestinians.

12. The Department of Palestinian Affairs oversees much of the management of Palestinian refugees in Jordan, which will be discussed in chapter 7.

13. Marshood (2010, 7) refers to Palestinians as "second class" refugees in comparison to those under the UNHCR mandate.

14. Citizenship in many Arab countries, including Jordan, is granted through paternal lineage. Thus a woman cannot pass on her citizenship or nationality to her children. This is a contentious point in Jordan, and civil society groups have been fighting to change this patriarchal law.

15. In 1995, in Libya, Qaddafi expelled Palestinians in order to, at least rhetorically, not acquiesce to Israel's desires to permanently remove Palestinians from Israel/Palestine.

16. Jordan's 1954 nationality law granted Jordanian nationality to all non-Jewish Palestinians who lived in Jordan (which included the West Bank) between December 20, 1949, and February 16, 1954. This 1954 law amended the 1928 nationality law and has been modified subsequently.

17. UN Secretary-General, "Note under General Assembly Resolution 2252 (ES-V) and Security Council Resolution 237 (1967)," https://digitallibrary.un.org/record/506880?ln=en.

18. From the 1920s to the 1940s, King Abdullah I embraced his own style of Arab nationalism that sought to unite Syria, Palestine, and Jordan. However, King Abdullah's commitment to Arab nationalism was greatly rhetorical. Over his thirty-year reign, Abdullah I maintained relations with the British, negotiated with Israel, and undermined Palestinian independence with the annexation of the West Bank and the proclamation of Jordanian guardianship over Palestine and Palestinians. Abdullah I was assassinated in 1951 in Jerusalem by a Palestinian, who, it is speculated, killed him because of his relations with Israel and annexation of the West Bank. See "Assassination of King Abdullah," *The Guardian*, July 21, 1951, https://www.theguardian.com/theguardian/1951/jul/21/fromthearchive.

19. Some Jordanians have been resentful of the aid given to Palestinian refugees, which has led to tensions (Abu-Odeh 1999, 137).

20. In 1994, Jordan and Israel signed a peace treaty, normalizing relations. This treaty recognized Jordan's safeguarding of the holy sites of Jerusalem, which would later be obliterated with US president Donald Trump's December 2017 declaration of Jerusalem as the capital of Israel. Many Palestinians consider the peace treaty between Jordan and Israel as strategic, but hurtful.

21. UNHCR, "Syria Emergency," https://www.unhcr.org/en-us/syria-emergency.html.

22. The "Arab Spring" (*al-Rabi' al-'Arabi*) at first seemed to herald Arab connections across state borders, but the "Arab Spring" was not just about or led by Arabs, and the transitions and

conflicts that ensued have not created Arab cohesion, nor have they been very "spring"-like. Drought, poverty, and internal migrations also fomented the uprising in Syria (Yassin-Kassab and Al-Shami 2016, 33).

23. ACAPS, "Jordanian Syrian Refugees: https://www.acaps.org/country/jordan/crisis/syrian-refugees.

24. Prima facie (meaning "at first appearance") status is used by the UNHCR to accommodate mass displacements quickly, as opposed to making individual determinations, which takes much more time (Rutinwa 2002). The UNHCR has not classified Syrians in any state as prima facie (Janmyr 2017; Chatty 2018, 230). However, through my interviews with experts in Jordan, as well as the research published by Lenner and Schmelter (2016) and Su (2013), we all found that prima facie has been used as a de facto practice. This ambiguity over prima facie and individual determination is yet another example of Jordan's complex and unclear system.

25. Further, within Azraq there is a prison in "village 5" where about 8,500 Syrians are detained. See Karen Laub and Alice Su, "Aid Groups: 8,500 Syrians Still Held in Jordanian No-Go Camp," *AP News*, January 30, 2018, https://apnews.com/5782dcbf32af4fb19f78de317717bd1b/Aid-groups:-8,500-Syrians-still-held-in-Jordanian-no-go-camp.

26. In the Gulf States, *kefala* binds workers to their employers by making the latter legally responsible for the former. It also makes a worker's stay conditional on having a work contract, putting workers in precarious situations that employers can and do exploit.

27. One expert I interviewed noted that monitoring was important just for knowledge about who was in Jordan and had no other intent, while another experts asserted that this practice was about security. While not a pointed theme in this book, it is relevant to note that this is a biopolitical practice of population control. Refugees are monitored in many ways, including tracking their caloric intake and bodily injuries to prove their persecution (Fassin 2012).

28. Areej Abuqudairi, "Syria's War Haunts Jordanian Border Town," *Al Jazeera*, May 11, 2015, https://www.aljazeera.com/news/2015/05/150510083150067.html.

29. Amnesty International, "Jordan: Authorities Must Allow Urgent Medical Care for Displaced Syrians in Rukban during COVID-19," May 7, 2020, https://www.amnesty.org/en/latest/news/2020/05/jordan-authorities-must-allow-urgent-medical-care-for-displaced-syrians-in-rukban-during-covid19/.

30. Many enter without legal documentation—birth certificates, marriage certificates, passports—making status determinations difficult.

31. UNHCR, "UNHCR Works through the Night to Register Syrians Reaching Jordan," July 10, 2014, https://www.unhcr.org/en-us/news/makingdifference/2014/7/53be89a49/unhcr-works-night-register-syrians-reaching-jordan.html.

32. The Jordanian Ministry of Labor maintains a list of professions and industries in which foreigners are prohibited from working. These include medicine; engineering; administrative, accounting, and clerical professions; telephone and warehouse employment; sales; education; hairdressing; decorating; fuel sales; electrical and mechanical occupations; security; driving; and construction work.

33. I stress *most* and *mostly* because there are Syrians whose needs are not being met, most notably those who do not have official papers and registration. Food insecurity is an issue too. The World Food Program—as of summer 2021—is running out of funds and may not be able to support Syrians with basic sustenance.

34. Syrians generally do not use the label "refugee" to refer to themselves because of its negative connotations.

35. Many Syrians I interviewed noted that they feared retribution by the Assad regime if they returned. Nevertheless, Syrians have been returning even though there is no guarantee of safety. They are doing so because they have suffered precarious living conditions, economic hardship, discrimination, loss of connection to family remaining in Syria, and a sense of not belonging in Jordan. See Saskia Baas, "The Real Reasons Why Syrians Return to Syria," *Refugees Deeply*, March 6, 2018, https://www.newsdeeply.com/refugees/community/2018/03/06/the-real-reasons-why-syrians-return-to-syria.

36. UNHCR, "Regional Resettlement Data," https://www.unhcr.org/en-au/regional-resettlement-data.html.

37. While the full costs of hosting refugees in Jordan has not been met by international donors and aid organizations, it is important to note that refugees have been a source of revenue for Jordan. International donors have provided humanitarian funds for refugee management, as well as for broader developmental projects. For example, Jordan received US$400 million between 2007 and 2009 to help support Iraqi refugees, and some of these funds went to developing schools, hospitals, and water and sewer pipes to benefit residents in the capital of Amman (El-Abed 2014, 95).

Chapter Four

1. A "homeland" is generally considered to be a place of origin in which people have ancestral (often ethnic or cultural) connections. A homeland's borders can be ambiguous, but a homeland is greatly territorial in that it has strong connotations about who is included and who is excluded. The term "homeland" is sometimes akin to the idea of country, state, or nation-state. The idea of the "nation-state" is sometimes used to refer to a territorial state, but a nation-state is the (theoretical) merging of a nation, which is a homogenous group of people who share a similar culture, history, and language, within an independent territorial state. Japan and Iceland are the only two modern independent states that seem to meet the criteria of a nation-state, but nevertheless, this term is used ubiquitously in academia and public discourse. As briefly noted earlier, I expressly reject the use of the term "nation-state" because it does not accurately represent the characteristics of the political ordering of the world.

2. Turks, Sunnis, Alawites, Greek Orthodox, Armenian Christians, Circassians, Chechens, Turcoman, Persians, and Assyrians all lived in this area under the Ottomans (Dawn and Finlayson 2010; Hathaway 2008, 3, 188).

3. Though there has been some debate about religious tolerance, it is generally agreed that the Ottomans were tolerant of non-Muslim minority groups, even though they applied a "dhimmi" tax to non-Muslims.

4. Some ideas in this paragraph stem from my 2014 *Cartographica* article but have been edited greatly.

5. Stemming directly from the declaration by ISIS/Daesh in 2014 that its establishment of a new Islamic state was a "fixing of the Sykes-Picot agreement," there has been renewed discussion of the relevance of the Sykes-Picot agreement and of imperial borders in SWANA more broadly.

6. *Sham* is the Arabic word for "north." *Bilad* translates as "our land, country, or homeland," but it can also indicate a community, village, or town (Hammer 2005, 73). *Al-Sham* literally translates as "to the left," which is "north" if one is in the Hejaz looking eastward, but is commonly used to refer to modern-day Syria (and even sometimes Damascus).

7. Ottomans divided *Bilad al-Sham* into the four provinces of Aleppo, Sham, Trablus al-Sharaq, and Raqqa (Raymond 1996, 122).

8. Many Arabs of North Africa, who were under the rule of the British and French, fought on the side of the Allies, often as conscripts. Some Arabs under Ottoman control fought with the Ottomans, while Hejazi Arabs led the Arab revolt against the Ottomans.

9. Arab nationalism is a layered ideology and practice that has evolved within different political and geopolitical contexts and has materialized in different forms (e.g., Hashemite, Nasserist, and Ba'athist). Like other identities, "Arab" does not have an essential core or singular origin. Instead, ideas and feelings of Arabness (*al-'uruba*) are created and facilitated by a shared culture, language, and historical experience, and by performances (Yassin-Kassab and Al-Shami 2016, 8; Shami 1996, 11; Dionigi 2017). Geography and territory, and the idea that Arabs originated from the Arabian Peninsula and migrated to create an expansive *al-Watan al-Arabi*, are central in the discourse (Chatelard 2010, 20–21, 38).

10. *Al-watan* in some contexts means a local territory or state-territory, but it also means "homeland" (Hammer 2005, 68). The term *qawmiyyia* is similar, referring to a "trans-territorial loyalty to an Arab cultural-linguistic unity" (Gershoni et al. 2011).

11. While some of my interviewees recognized the connective effects of Arabic, many noted that dialects can be so different as to be incomprehensible. Religion was on occasion noted as a connecting factor too. Islam and its teachings were explicitly mentioned, but so too was religion in the more historical sense of being central to that area of the world. That Arabs were generous and welcoming, as a cultural practice, was also mentioned by some Palestinians as a shared trait that connected them across borders.

12. There is a lot of linguistic and religious diversity in Syria. In addition to Arabs, there are Circassians, Turkmen, and Kurds, for example.

13. Some Syrians viewed Islam as a major reason for their sense of connection to Jordan. While Jordanians and Syrians are not all Muslim, Islam is the dominant religion in both states, and the majority of my interviewees self-identified as Muslim or mentioned their religious practices during our interviews.

14. The Cairo Declaration on Human Rights in Islam (1993), Rights of the Child in Islam (2005), the Arab Declaration for International Migration (2006), and the Arab Observatory for International Migration are all examples of international instruments that protect human rights, including mobility within an Arab and/or Islamic frame. The Arab Charter on Human Rights (completed in 1994, edited in 2004, and ratified in 2008) protects against refoulement, the human right to secure basic needs, access to the courts, freedom of movement, and freedom from arbitrary detention (Bidinger et al. 2014). The 1981 Universal Islamic Declaration of Human Rights lists twenty-three rights, of which the ninth is "right to asylum." In January 2004, Egypt and Sudan signed the Four Freedom Agreement, which grants citizens the right to move, reside, work, and own property in either country. Recent diplomatic rifts, however, have threatened this agreement.

15. This declaration was signed after a series of four meetings on asylum and refugee law between 1984 and 1992, but it was never ratified and thus never formally applied. See "Declaration on the Protection of Refugees and Displaced Persons in the Arab World," https://www.refworld.org/docid/452675944.html.

16. These other legal instruments include the Universal Declaration of Human Rights, the International Covenant on Civil and Political Rights, and the International Covenant of Economic, Social and Cultural Rights.

17. "Declaration on the Protection of Refugees and Displaced Persons in the Arab World," https://www.refworld.org/docid/452675944.html.

18. While offering residency rights, the authors of the Casablanca Protocol were careful to assert that such rights would not relinquish the Palestinian "right of return" to the land that Israel occupied in 1948.

19. UNHCR, "The 1951 Refugee Convention," http://www.unhcr.org/en-us/protection/basic/3b73b0d63/states-parties-1951-convention-its-1967-protocol.html.

20. Jordan's 1952 constitution labels the country an Arab state. Other examples in which Jordanian policies have been framed by Arab nationalism include (1) that the Jordanian army has been officially referred to as the "Arab army" and the "Arab Legion"; (2) that Jordan did not support the US-led allies in the 1991 Gulf War, largely because popular Arab sentiment saw the US invasion as an encroachment on fellow Arab lands; and (3) that nearly all Jordanians speak Arabic and identify as Arab.

21. Reciprocity exists between these states and Jordanian passport holders, meaning that Jordanians have visa-free access in several neighboring Arab states too. See Jordan Ministry of Interior, "Restricted and Non-restricted Countries," http://international.visitjordan.com/general information/entryintojordan.aspx.

Chapter Five

1. The border was open in the other direction, as Israel allowed Palestinians to migrate to Jordan and West Bank goods were allowed to be sold to in Jordan (Abu-Odeh 1999, 139).

2. Jordan officially relinquished control of the West Bank in 1988 but lost administrative and military control when Israel invaded and occupied the West Bank in 1967.

Chapter Six

1. Parts of this paragraph are echoed in my *Geopolitics* article from 2016.

2. For thorough discussions on the evolving geographies of Palestine and Historical Palestine, see Hammer (2005), Khalidi (1997), Peteet (1995), and Davis (2011); and for Israel and the Holy land, see Havrelock (2011).

3. Two territories of the Syrian mandate have been annexed. The region of Alexandretta was lost to Turkey and the Golan Heights to Israel.

4. The Ba'ath party came to power in 1963 and remains. Ba'athism is, in theory, nationalist (as in Syria and Iraq) as well as pan-Arabist and quasi-socialist.

Chapter Seven

1. UNHCR (2019); UNRWA, "Where We Work: Jordan," https://www.unrwa.org/where-we-work/jordan; UNHCR, "What Is a Refugee Camp? Definitions and Statistics," https://www.unrefugees.org/refugee-facts/camps/.

2. There are countless other refugee and IRR spaces that have been built or managed by the Jordanian government, INGOs, NGO, and refugees (Sanyal 2017). Registration and transit centers, run by the state and IRR organizations, are scattered in the north and around Amman particularly. There are refugee community centers and medical and mental health care clinics in towns and cities. It is quite common for refugees to establish restaurants and other businesses

as well. Border crossings, berm spaces, and militarized zones have been also become part of the broad refugee landscape of Jordan.

3. Agamben's theories of sovereign power, the ubiquity of the state of exception today, spaces of exception, and "bare life" have become central, if not obligatory, to scholarship on human rights, political rights, refugees, and camps (Abourahme 2015, 201; Singh 2020). While I recognize Agamben's influence, his work is not central to my analysis, as it is quite abstract and ahistorical, lacks specificity, does not apply well to Palestinian camps (as I note in the main text), and too often dismisses the agency of refugees, who I assert are resilient and strong in the face of their many challenges.

4. Camp residents do not own the land in the camps. For the most part, the land is rented to the Jordanian government by private Jordanian citizens.

5. For discussions on the politics of UNRWA see Farah (2012), Abourahme (2014), and Robson (2017).

6. UNRWA, "Jerash Camp," https://www.unrwa.org/where-we-work/jordan/jerash-camp.

7. The official Palestinian leader of the camp, who works with the DPA, provided me with this data during our interview in March 2018.

8. Many refugees from the Gaza Strip were displaced a second time when they sought refuge in Jordan, as they had already been displaced in 1948 from other areas of Israel/Palestine to the Gaza Strip. These people often do not consider themselves as being from the Gaza Strip.

9. UNRWA, "Jerash Camp," https://www.unrwa.org/where-we-work/jordan/jerash-camp.

10. The Jordanian government introduced new socioeconomic programs and policies in the mid-1990s to help facilitate connections between the camps and the urban localities surrounding them (Hanafi 2010, 11).

11. In some ways, applying the term "dasymetric" to camps is akin to a topology. As Abourahme (2014, 212) explains, boundaries between a city and a camp can be fuzzy and can create a kind of "topological folding" and "spillover." I prefer the former term, though, because of its specificity in examining spatial relations, whereas "topological" is much broader and stems from mathematics.

12. In these camps, 97.5 percent of residents are Palestinian. Those 2.5 percent who are not are in the camp mostly due to intermarriage.

13. Many Palestinians have never lived in camps. Palestinians outside of camps live in houses and apartment complexes that resemble non-refugee households much more than they resemble Palestinian households inside the camps (Tiltnes and Zhang 2013, 47).

14. Jordan was in a state of martial law from the outbreak of the war in 1970 until 1989, which was one year after Jordan's disengagement from the West Bank.

15. Jordan had initially treated PRS the same as Syrian nationals, but the government began to refuse PRS entry in early 2012. Then, in 2013, Jordan formalized a policy of non-entry for PRS (Francis 2015, 23; Hassan 2018). UNRWA reported that 13,836 PRS had sought their support in Jordan as of April 2014, but there are likely thousands more who are in Jordan but have not sought UNRWA support. Palestinians are UNRWA-mandated refugees. They are not under the auspices of UNHCR and are thus subject to different policies and aid structures than Syrian nationals. As of April 2014, almost two hundred Palestinians from Syria were detained in Cyber City, the government-created camp/detention center, along with two hundred non-Palestinians, most of whom had familial relations to a PRS, either through marriage or through being single parents to Palestinian children. Palestinians in Cyber City can leave every few weeks to visit family they might have in Jordan for a maximum period of forty-eight hours (Hassan 2018).

Unlike Syrian refugees, these Palestinians are not eligible for "bail out" by a Jordanian sponsor (Bidinger et al. 2014, 66), so the only other way they can leave the camp is to return to Syria. The 3RP report discussed above notes that PRS *should* be treated the same as all other Syrian refugees, while stressing that their status as refugees in Jordan should not affect their "right of return" in accordance with UN Resolution 194. I did not interview any PRS, which is why this information is relegated to a note.

16. Stephanie Ott, "Syrians at Za'atari Camp: 'We Can't Live Here Forever,'" *Al Jazeera*, October 24, 2015, https://www.aljazeera.com/news/2015/10/24/syrians-at-zaatari-camp-we-cant-live-here-forever.

17. Camp residents can get passes to leave for short vacations outside and for medical issues and appointments.

Conclusion

1. UNHCR, "Regional Resettlement Data," https://www.unhcr.org/en-au/regional-resettlement-data.html.

2. Agamben (1998,126, 134) asserts that refugees and refugee camps exist because the international community of states is incapable of dealing with statelessness and mass displacements. Refugees and stateless people, Agamben argues, are included in the political order of the world only through their exclusion from normal rights, which are generally derived from states. Reflecting on Arendt, he stresses that "the refugee" should be the exemplar of rights, but instead refugees demonstrate a crisis of rights that happens when a person is no longer part of a state.

3. The UN's 2018 refugee budget for Africa, the "Middle East" and North Africa, Asia and the Pacific, and the Americas totaled US$4.352 billion; Global North countries spent $20 billion to deter migrants (Hathaway 2018, 593).

Bibliography

Abourahme, Nasser. 2015. "Assembling and Spilling-Over: Towards an 'Ethnography of Cement' in a Palestinian Refugee Camp." *International Journal of Urban and Regional Research* 39: 200–217.

Abu-Odeh, Adnan. 1999. *Jordanians, Palestinians and the Hashemite Kingdom in the Middle East Peace Process*. Washington, DC: US Institute of Peace.

ACAPS. 2022. "Jordan Syrian Refugees." https://www.acaps.org/country/jordan/crisis/syrian-refugees.

Achilli, Luigi. 2014. "Disengagement from Politics: Nationalism, Political Identity, and the Everyday in a Palestinian Refugee Camp in Jordan." *Critique of Anthropology* 34: 234–57.

———. February 2015. *Syrian Refugees in Jordan: A Reality Check*. Fiesole, Italy: Migration Policy Centre, European University Institute.

Adelson, Roger. 1995. *London and the Invention of the Middle East: Money, Power, and War, 1902–1922*. New Haven, CT: Yale University Press.

Adepoju, Aderanti, Alistair Boulton, and Mariah Levin. 2010. "Promoting Integration through Mobility: Free Movement under ECOWAS." *Refugee Survey Quarterly* 29 (3): 120 44.

AFP. 2020. "Syria Death Toll Tops 380,000 in Almost Nine-Year War: Monitor." April 1, 2020. https://www.france24.com/en/20200104-syria-death-toll-tops-380-000-in-almost-nine-year-war-monitor.

Agamben, Giorgio. 1998. *Homo Sacer: Sovereign Power and Bare Life*. Stanford, CA: Stanford University Press.

———. 2005. *State of Exception*. Chicago: University of Chicago Press.

Agier, Michel. 2002. "Between War and City: Towards an Urban Anthropology of Refugee Camps." *Ethnography* 3 (3): 37–341.

Agnew, John. 1994. "The Territorial Trap: The Geographic Assumptions of International Relations Theory." *Review of International Political Economy* 1 (1): 53–80.

———. 2019. "The Asymmetric Border: The United States' Place in the World and the Refugee Panic of 2018." *Geographical Review* 109 (4): 507–26.

al-Shoubaki, Hind. 2017. "The Temporary City: The Transformation of Refugee Camps from Fields of Tents to Permanent Cities." *European Planning Studies* 7:5–15.

Aleinikoff, T. Alexander. 1995. "State-Centered Refugee Law: From Resettlement to Containment." In *Mistrusting Refugees*, edited by E. Valentine Daniel and John Chr. Knudsen, 257–78. Berkeley, CA: University of California Press.

Alnsour, Jamal, and Julia Meaton. 2014. "Housing Conditions in Palestinian Refugee Camps, Jordan." *Cities* 36:65–73.

Amin, Samir. 1978. *The Arab Nation*. London: Zed.

Anderson, Betty. 2005. *Nationalist Voices in Jordan: The Street and the State*. Austin, TX: University of Texas Press.

Anderson, Stuart. 2022. "Refugees Make America Better Off." *Forbes*, May 26, 2022. https://www.forbes.com/sites/stuartanderson/2022/05/26/refugees-make-america-better-off/.

Antonius, George. 1965. *The Arab Awakening: The Story of the Arab National Movement*. New York: Capricorn Books.

Antonsich, Marco. 2010. "Searching for Belonging: An Analytical Framework." *Geography Compass* 4 (6): 644–59.

Anzaldúa, Gloria. 1999. *Borderlands/La Frontera*. 2nd ed. San Francisco: Aunt Lute Books.

Appadurai, Arjun. 1996a. *Modernity at Large*. Minneapolis: University of Minnesota Press.

———. 1996b. "Sovereignty without Territoriality." In *The Geography of Identity*, edited by Patricia Yaeger, 40–58. Ann Arbor, MI: University of Michigan Press.

Arendt, Hannah. 1951. *The Origins of Totalitarianism*. Orlando, FL: Harvest Books.

Ballvé, Teo. 2012. "Everyday State Formation: Territory, Decentralization, and the Narco Landgrab in Colombia." *Environment and Planning D: Society and Space* 30:603–22.

Barakat, Rana. 2013. "The Right to Wait: Exile, Home and Return " In *Seeking Palestine*, edited by Penny Johnson and Raja Shehadeh, 135–47. Northampton, MA: Olive Tree Press.

Barnett, Thomas. 2003. "The Pentagon's New Map." *Esquire*, March 2003. Republished at https://www.esquire.com/news-politics/a1546/thomas-barnett-iraq-war-primer/.

Basset, Abdel, and Fooaz Yaoob al-Moony. 2016. *Jordanian Perceptions of Syrian Refugees*. Irbid, Jordan: Yarmouk University.

Berman, Chantal. 2012. "An Uncommon Burden: Aid, Resettlement, and Refugee Policy in Syria," in *Transatlantic Cooperation on Protracted Displacement: Urgent Need and Unique Opportunity*, edited by John Calabrese and Jean-Luc Marret. Washington, DC: Middle East Institute Press.

Bernstein, Jesse, and Moses Chrispus Okello. 2007. "To Be or Not to Be: Urban Refugees in Kampala." *Refuge* 24 (1): 46–56.

Betts, Alexander. 2013. "Regime Complexity and International Organizations: UNHCR as a Challenged Institution." *Global Governance* 19 (1): 69–81.

Bhabha, Homi K. 1994. *The Location of Culture*. New York: Routledge.

———. 2015. "Foreword." In *Debating Cultural Hybridity: Multicultural Identities and the Politics of Anti-racism*, edited by Pnina Werbner and Tariq Modood. London: Zed Books.

Bidinger, Sarah, Aaron Lang, Danielle Hites, Yoana Kuzmova, Elena Noureddine, and Susan M. Akram. 2014. "Protecting Syrian Refugees: Laws, Policies, and Global Responsibility Sharing." Report, Boston University School of Law. https://www.bu.edu/law/files/2015/08/syrianrefugees.pdf.

Billé, Franck. 2020. "Voluminous: An Introduction." In *Voluminous States: Sovereignty, Materiality, and the Territorial Imagination*, edited by Franck Billé, 1–35. Durham, NC: Duke University Press.

Bloch, Alice. 2020. "Reflections and Directions for Research in Refugee Studies." *Ethnic and Racial Studies* 43 (3): 436–59.

Bose, Pablo S. 2020. "The Shifting Landscape of International Resettlement: Canada, the US and Syrian Refugees." *Geopolitics*: 27 (2): 1–27.

Brand, Laurie. 1995. "Palestinians and Jordanians: A Crisis of Identity." *Journal of Palestine Studies* 24 (4): 46–61.

Brown, Carl L. 1996. *Imperial Legacy: The Ottoman Imprint on the Balkans and the Middle East.* New York: Columbia University Press.

Bryan, Joe. 2012. "Rethinking Territory: Social Justice and Neoliberalism in Latin America's Territorial Turn." *Geography Compass* 6 (4): 215–26.

Burgis, Michelle. 2009. "Faith in the State? Traditions of Territoriality, International Law and the Emergence of Modern Arab Statehood." *Journal of the History of International Law* 11:37–79.

Cabot, Heath. 2016. "'Refugee Voices':Tragedy, Ghosts, and the Anthropology of Not Knowing." *Journal of Contemporary Ethnography* 45 (6): 645–72.

Cain, Trudie, Carina Meares, and Christine Read. 2015. "Home and Beyond in Aotearoa: the Affective Dimensions of Migration for South African Migrants." *Gender, Place and Culture* 22 (8): 1141–57.

Carter, Sean. 2005. "The Geopolitics of Diaspora." *Area* 37 (1): 54–63.

Chakrabarty, Dipesh. 2000. *Provincializing Europe.* Princeton, NJ: Princeton University Press.

Chatelard, Geraldine. 2010a. "What Visibility Conceals: Re-embedding Refugee Migration from Iraq." In *Dispossession and Displacement: Forced Migration in the Middle East and North Africa*, edited by Dawn Chatty and Bill Finlayson, 17–44. Oxford: Oxford University Press.

———. 2010b. "Jordan: A Refugee Haven." Migration Policy Institute, August 31, 2010. https://www.migrationpolicy.org/article/jordan-refugee-haven.

Chatty, Dawn. 2010. *Displacement and Dispossession in the Modern Middle East.* New York: Cambridge University Press.

———. 2016. "Refugee Voices: Exploring the Border Zones between States and State Bureaucracies." *Refuge* 32 (1): 3–6.

———. 2017. "The Duty to be Generous (Karam): Alternatives to Rights-Based Asylum in the Middle East." *Journal of the British Academy* 5:177–99.

———. 2018. *Syria: The Making and Unmaking of a Refuge State.* New York: Oxford University Press.

Chatty, Dawn, and Nisrine Mansour. 2011. "Unlocking Protracted Displacement: An Iraqi Case Study." *Refugee Survey Quarterly* 30 (4): 50–83.

Chimni, B. S. 2019. "Global Compact on Refugees: One Step Forward, Two Steps Back." *International Journal of Refugee Law* 30 (4): 630–34.

Clare, Nick, Victoria Habermehl, and Liz Mason-Deese. 2018. "Territories in Contestation: Relational Power in Latin America." *Territory, Politics, Governance* 6 (3): 302–21.

Clayton, Jonathan, and Hereward Holland. Dec 30 2015. "Over One Million Sea Arrivals Reach Europe in 2015." UNHCR, December 30, 2015. https://www.unhcr.org/en-us/news/latest/2015/12/5683d0b56/million-sea-arrivals-reach-europe-2015.html.

Clutterbuck, Martin, Yara Hussein, Mazen Mansour, and Monica Rispo. 2021. "Alternative Protection in Jordan and Lebanon: The Role of Legal Aid." *Forced Migration Review*. https://www.fmreview.org/issue67/clutterbuck-hussein-mansour-rispo.

Cockburn, Cynthia. 2007. *From Where We Stand: War, Women's Activism, and Feminist Analysis.* New York: Zed Books.

Coddington, Kate. 2018a. "Settler Colonial Territorial Imaginaries: Maritime Mobilities and the "Tow-Backs" of Asylum Seekers." In *Terra beyond Territory*, edited by Kimberly Peters, Philip Steinberg, and Elain Stratford, 185–201. Lanham, MD: Rowman & Littlefield.

———. 2018b. "Landscapes of Refugee Protection." *Transactions of the British Institute of Geographers* 43:326–40.

Cole, Georgia. 2021. "Pluralising Geographies of Refuge." *Progress in Human Geography* 45 (1): 88–110.

Collins, Francis L. 2022. "Geographies of Migration II: Decolonising Migration Studies." *Progress in Human Geography* 46 (5): 1241–51.

Collyer, Michael, and Russell King. 2015. "Producing Transnational Space: International Migration and the Extra-territorial Reach of State Power." *Progress in Human Geography* 39 (2): 185–204.

Conlon, Deirdre, and Nancy Hiemstra. 2017. "Introduction." In *Intimate Economies of Immigrant Detention*, edited by Deirdre Conlon and Nancy Hiemstra, 1–12. New York: Routledge.

Cresswell, Tim. 2013. *Geographic Thought: A Critical Introduction.* Malden, MA: Wiley-Blackwell.

Culcasi, Karen. 2010. "Constructing and Naturalizing the Middle East." *Geographical Review* 100 (4): 583–97.

———. 2011. "Cartographies of Supranationalism: Creating and Silencing Territories in the 'Arab Homeland.'" *Political Geography* 30:417–28.

———. 2012. "Mapping the Middle East from Within: (Counter) Cartographies of an Imperialist Construction" *Antipode* 44 (4): 1099–118.

———. 2014. "Disordered Ordering: Mapping the Division of the Ottoman Empire." *Cartographica* 49 (1): 2–17.

———. 2016. "Warm Nationalism: Mapping and Imagining the Jordanian Nation." *Political Geography* 54:7–20.

———. 2017. "Imagining Arab Homeland and Palestine." In *Scaling Identities: Nationalism and Territoriality*, edited by Guntram Herb and David Kaplan, 137–55. Lanham, MD: Rowman & Littlefield.

———. 2019. "'We Are Women and Men Now': Intimate Spaces and Coping Labour for Syrian Women Refugees in Jordan." *Transactions of the British Institute of Geographers* 44:463–78.

Culcasi, Karen, Emily Skop, and Cynthia Gorman. 2019. "Contemporary Refugee-Border Dynamics and the Legacies of the 1919 Paris Peace Conference." *Geographical Review* 109: 469–86.

Dagtas, Secil. 2017. "Whose Misafirs? Negotiating Difference along the Turkish-Syrian Border." *International Journal of Middle East Studies* 49:661–79.

Dahlman, Carl. 2009. "Territory." In *Key Concepts in Political Geography*, edited by Carolyn Gallaher, Carl Dahlman, Mary Gilmartin, and Alison Mountz. Thousand Oaks, CA: Sage.

Darling, Jonathan. 2017. "Forced Migration and the City: Irregularity, Informality, and the Politics of Presence." *Progress in Human Geography* 41 (2): 178–98.

Darwish, Mahmoud. 2010. *Journal of Ordinary Grief.* Translated by Ibrahim Muhawi. New York: Archipelago Books.

Davis, Rochelle. 2011. *Palestinian Village Histories: Geographies of the Displaced.* Stanford, CA: Stanford University Press.

Davis, Rochelle, and Abbie Taylor. 2013. *Syrian Refugees in Jordan and Lebanon: A Snapshot from Summer 2013*. Center for Contemporary Arab Studies, Institute for the Study of International Migration, Georgetown University.

Dawn, Ernest. 1991. "The Origins of Arab Nationalism." In *The Origins of Arab Nationalism*, edited by Rashid Khalidi, Muhammad Muslih, and Reeva Simon, 3–30. New York: Columbia University Press.

De Genova, Nicholas. 2018. "Rebordering 'the People': Notes on Theorizing Populism." *South Atlantic Quarterly* 117 (2): 357–74.

De Genova, Nicholas, Glenda Garelli, and Martina Tazzioli. 2018. "Autonomy of Asylum? The Autonomy of Migration Undoing the Refugee Crisis Script." *South Atlantic Quarterly* 117:239–65.

de Vet, Annelys, ed. 2007. *Subjective Atlas of Palestine*. Rotterdam, Netherlands: 010 Publishers.

Debarbieux, Bernard. 2019. *Social Imaginaries of Space*. Cheltenham, UK: Edward Elgar.

Delaney, David. 2005. *Territory: A Short Introduction*. Malden, MA: Blackwell.

Delaney, David, and Päivi Rannila. 2021. "Scopic Relations as Spatial Relations." *Progress in Human Geography* 45 (4): 704–19.

Deleuze, Gilles, and Felix Guattari. 1987. *A Thousand Plateaus: Capitalism and Schizophrenia*. Minneapolis: University of Minnesota Press.

dell'Agnese, Elena. 2013. "The Political Challenge of Relational Territory." In *Spatial Politics: Essays for Doreen Massey*, edited by David Featherstone and Joe Painter, 116–24. Malden, MA: Wiley-Blackwell.

Dempsey, Kara E. 2018. "Negotiated Positionalities and Ethical Considerations of Fieldwork on Migration: Interviewing the Interviewer." *ACME* 17:88–108.

Dionigi, Filippo. 2017. "Rethinking Borders: The Case of the Syrian Refugee Crisis in Lebanon." In *Refugees and Displacement in the Middle East*, edited by Marc Lynch and Laurie Brand, 22–29. Washington, DC: Project on Middle East Political Science.

Dorai, Mohamed Kamel. 2002. "The Meaning of Homeland for the Palestinian Diaspora: Revival and Transformation." In *New Approaches to Migration? Transnational Communities and the Transformation of Home*, edited by Nadje Al-Ali and Khalid Koser. London: Routledge.

Doucet, Lyse. 2012. "Jordan's Desert Camp for Syrian Refugees. *BBC News*, July 30, 2010. https://www.bbc.com/news/world-middle-east-19042686.

Ehrkamp, Patricia. 2017a. "Geographies of Migration I: Refugees." *Progress in Human Geography* 41 (6): 813–22.

Eid, Omar Abdullah Al-Haj. 2019. "Writing on Tents and Caravans in al Za'atari Syrian Refugee Camp of Mafraq, Jordan: A Sociolinguistic Analysis." *Humanities & Social Sciences Reviews* 7 (5): 352–63.

El-Abed, Oroub. 2005. "Immobile Palestinians: Ongoing Plight of Gazans in Jordan." *Forced Migration Report* 26:7–8.

———. 2014. "The Discourse of Guesthood: Forced Migrants in Jordan." In *Managing Muslim Mobilities*, edited by Anita H. Fábos and Riina Isotalo, 81–100. New York: Palgrave Macmillan.

El Dardiry, Giulia. 2017. "'People Eat People': The Influence of Socioeconomic Conditions on Experiences of Displacement in Jordan." *International Journal of Middle East Studies* 49: 01–19.

Elden, Stuart. 2005. "Missing the Point: Globalization, Deterritorialization and the Space of the World." *Transactions of the Institute of British Geographers* 30:8–19.

———. 2009. *Terror and Territory: The Spatial Extent of Sovereignty*. Minneapolis: University of Minnesota Press.

———. 2013a. *The Birth of Territory*. Chicago: University of Chicago Press

———. 2013b. "The Significance of Territory: Review of Jean Gottmann." *Geographica Helvetica* 68:65–68.

Enloe, Cynthia. 2000. *Bananas, Beaches and Bases: Making Feminist Sense of International Politics*. 2nd ed. Berkeley: University of California Press.

Eurostat. 2015. "*EU Population Up to 508.2 Million at 1 January 2015*." News release, July 10, 2015. https://ec.europa.eu/eurostat/documents/2995521/6903510/3-10072015-AP-EN.pdf/.

Fábos, Anita H., and Riina Isotalo. 2014. "Introduction: Managing Muslim Mobilities—A Conceptual Framework." In *Managing Muslim Mobilities*, edited by Anita H. Fábos and Riina Isotalo. New York: Palgrave Macmillan.

Fábos, Anita, and Gaim Kibreab. 2007. "Urban Refugees: Introduction." *Refuge* 24 (1): 3–10.

Farah, Randa. 2008. "Refugee Camps in the Palestinian and Sahrawi National Liberation Movements: A Comparative Perspective." *Journal of Palestine Studies* 38 (2): 76–93.

———. 2012. "Keeping an Eye on UNRWA." Policy brief, January 25, 2012. *al-Shabaka*. https://al-shabaka.org/briefs/keeping-eye-unrwa/.

Fassin, Didier. 2012. *Humanitarian Reason: A Moral History of the Present*. Berkeley: University of California Press.

Featherstone, David, and Joe Painter, eds. 2013. *Spatial Politics: Essays for Doreen Massey*. Malden, MA: Wiley-Blackwell.

Feldman, Ilana. 2015. "What Is a Camp? Legitimate Refugee Lives in Spaces of Long-Term Displacement." *Geoforum* 66:244–52.

———. 2016. "Punctuated Humanitarianism: Palestinian Life between the Catastrophic and the Cruddy." *International Journal of Middle East Studies* 48:372–76.

Ferraz de Oliveira, António. 2021. "Territory and Theory in Political Geography, c.1970s–90s: Jean Gottmann's *The Significance of Territory*." *Territory, Politics, Governance* 9:553–70.

Field, Robin, and Parmita Kapadia, eds. 2011. *Transforming Diaspora: Communities beyond National Boundaries*. Lanham, MD: Fairleigh Dickinson University and Rowman & Littlefield.

Foucault, Michel. 1986. "Of Other Spaces." Translated by Jay Miskowiec. *Diacritics* 16 (1): 22–27.

Francis, Alexandra. 2015. *Jordan's Refugee Crisis*. Carnegie Endowment for International Peace. https://carnegieendowment.org/2015/09/21/jordan-s-refugee-crisis-pub-61338.

Frohlich, Christiane, and Matthew R. Stevens. 2015. "Trapped in Refuge: The Syrian Crisis in Jordan Worsens." *Middle East Research and Information Project*. https://merip.org/2015/03/trapped-in-refuge/.

Fruchter-Ronen, Iris. 2013. "The Palestinian Issue as Constructed in Jordanian School Textbooks, 1964–94: Changes in the National Narrative." *Middle Eastern Studies* 49 (2): 280–95.

Gabiam, Nell. 2016. "Humanitarianism, Development, and Security in the 21st Century: Lessons from the Syrian Refugee Crisis." *International Journal of Middle East Studies* 48:382–86.

Gabiam, Nell, and Elena Fiddian-Qasmiyeh. 2017. "Palestinians and the Arab Uprisings: Political Activism and Narratives of Home, Homeland, and Home-Camp." *Journal of Ethnic and Migration Studies* 5:731–48.

Gandolfo, Luisa. 2012. *Palestinians in Jordan: The Politics of Identity*. New York: IB Tauris.

Gatrell, Peter, Anindita Ghoshal, Katarzyna Nowak, and Alex Dowdall. 2021. "Reckoning with Refugeedom: Refugee Voices in Modern History." *Social History* 46 (1): 70–95.

Gatter, Melissa. 2018. "Rethinking the Lessons from Za'atari Refugee Camp." *Forced Migration Review* 57:22–24.

Gelvin, James. 2018. *The Middle East: What Everyone Needs to Know*. New York: Oxford University Press.

Glăveanu, V., and T. Zittoun. 2017. *The Future of Imagination in Sociocultural Research. In Handbook of Imagination and Culture*. Oxford: Oxford University Press.

Goddard, Stacie. 2010. *Indivisible Territory and the Politics of Legitimacy: Jerusalem and Northern Ireland*. New York: Cambridge University Press.

Godin, Marie, and Giorgia Doná. 2016. "'Refugee Voices,' New Social Media and Politics of Representation: Young Congolese in the Diaspora and Beyond." *Refuge: Canada's Journal on Refugees* 32 (1): 60–71.

Gorman, Anthony, and Sossie Kasbarian, eds. 2015. *Diasporas of the Modern Middle East: Contextualizing Community*. Edinburgh: Edinburgh University Press.

Gorman, Cynthia, and Karen Culcasi. 2021. "Invasion and Colonization: Islamophobia and Anti-refugee Sentiment in West Virginia." *Environment and Planning C: Politics and Space* 39 (1): 168–83.

Gottmann, Jean. 1973. *The Significance of Territory*. Charlottesville, VA: University Press of Virginia.

Halvorsen, Sam. 2019. "Decolonising Territory: Dialogues with Latin American Knowledges and Grassroots Strategies." *Progress in Human Geography* 43 (5): 790–814.

Hamdan, Ali Nehme. 2016. "Breaker of Barriers? Notes on the Geopolitics of the Islamic State in Iraq and Sham." *Geopolitics* 21 (3): 605–27.

Hammer, Juliane. 2005. *Palestinians Born in Exile: Diaspora and the Search for a Homeland*. Austin: University of Texas Press.

Hanafi, Sari. 2010. *Governing Palestinian Refugee Camps in the Arab East: Governmentalities in Search of Legitimacy*. Beirut: American University of Beirut.

———. 2014. "Forced Migration in the Middle East and North Africa." In *Refugee and Forced Migration Studies*, edited by Elena Fiddian-Qasmiyeh, Gil Loescher, Katy Long, and Nando Sigona, 585–98. Oxford: Oxford University Press.

Hathaway, James. 2018. "The Global Cop-Out on Refugees." *International Journal of Refugee Law* 30 (4): 591–604.

Hathaway, Jane. 2008. *The Arab Lands under Ottoman Rule, 1516–1800*. Harlow, UK: Pearson.

Havrelock, Rachel. 2011. *River Jordan: The Mythology of a Dividing Line*. Chicago: University of Chicago Press.

Hawthorne, Camilla. 2019. "Black Matters Are Spatial Matters: Black Geographies for the Twenty-First Century." *Geography Compass* 13 (11): e12468. https://doi.org/10.1111/gec3.12468.

Hayes-Conroy, Allison. 2018. "Somatic Sovereignty: Body as Territory in Colombia's Legión del Afecto." *Annals of the American Association of Geographers* 108 (5): 1298–312.

Heffernan, Michael. 1995. "The Spoils of War: the Societe de Geographie de Paris and the French Empire, 1914–1919." In *Geography and Imperialism, 1820–1940*, edited by Morgan Bell, Robin Butlin, and Michael Heffernan, 221–64. New York: St. Martin's Press.

Hilal, Leila, and Dr. Shahira Samy. 2008. "Asylum and Migration in the Mashrek: Asylum and Migration Country Fact Sheet Jordan." Euro-Mediterranean Human Rights Network. https://idcoalition.org/wp-content/uploads/2009/09/factsheet_jordan_en1.pdf.

Hoffmann, Sophia. 2016a. "International Humanitarian Agencies and Iraqi Migration in Preconflict Syria." *International Journal of Middle East Studies* 48:339–55.

———. 2016b. *Iraqi Migrants in Syria: The Crisis before the Storm*. Syracuse, NY: Syracuse University Press.

———. 2017. "Humanitarian Security in Jordan's Azraq Camp." *Security Dialogue* 48 (2): 97–112.

Holmes, Seth, and Heide Castañeda. 2016. "Representing the 'European Refugee Crisis' in Germany and Beyond: Deservingness and Difference, Life and Death." *American Ethnologist* 43:1–13.

Horn, Denise M, and Serena Parekh. 2018. "Introduction to 'Displacement.'" *Signs* 43 (3): 503–14.

Howden, Daniel, Hannah Patchett, and Charlotte Alfred. 2017. "The Compact Experiment: Push for Refugee Jobs Confronts Reality of Jordan and Lebanon." *Refugees Deeply*. https://s3.amazonaws.com/newsdeeply/public/quarterly3/RD+Quarterly+-+The+Compact+Experiment.pdf.

Hyndman, Jennifer. 2007. "Feminist Geopolitics Revisited: Body Counts in Iraq." *Professional Geographer* 59 (1): 35–46.

———. 2009. "Second Class Immigrants or First Class Protextion? Resettling Refugees to Canada." In *Resettled and Included? Employment Integration of Refugees*, edited by P. Bevelander, M. Hogstrom, and S. Ronnqvist, 247–65. Malmö, Sweden: Malmö University.

International Labour Organization, Regional Office for Arab States. 2015. "*Access to Work for Syrian Refugees in Jordan: A Discussion Paper on Labour and Refugee Laws and Policies*." https://www.ilo.org/wcmsp5/groups/public/---arabstates/---ro-beirut/documents/publication/wcms_357950.pdf.

International Labour Organization, Regional Office for Arab States. 2017. "*Work Permits and Employment of Syrian Refugees in Jordan*." https://www.ilo.org/wcmsp5/groups/public/---arabstates/---ro-beirut/documents/publication/wcms_559151.pdf.

Isotalo, Riina. 2014. "Fear of Palestinization: Managing Refugees in the Middle East." In *Managing Muslim Mobilities*, edited by Anita H. Fábos and Riina Isotalo, 59–80. New York: Palgrave Macmillan.

Jaeger, Gilbert. 2001. "On the History of the International Protection of Refugees." *International Review of the Red Cross* 83 (843): 727–37.

Janmyr, Maja. 2017. "No Country of Asylum: 'Legitimizing' Lebanon's Rejection of the 1951 Refugee Convention." *International Journal of Refugee Law* 29 (3): 438–65.

Jeffrey, Alex. 2020. "Cultural Geographies of the State and Nation." In *Handbook on the Changing Geographies of the State: New Spaces of Geopolitics*, edited by Sami Moisio, Natalie Koch, Andrew E. G. Jonas, Christopher Lizotte, and Juho Luukkonen, 33–45. Cheltenham, UK: Edward Elgar.

Joffé, George. 2017. "States and Caliphates." *Geopolitics* 23 (3): 505–24.

Johnson, Penny, and Raja Shehadeh, eds. 2013. *Seeking Palestine: New Palestinian Writing on Exile and Home*. Northampton, MA: Olive Tree Press.

Johnson, Robert. 2018. "The de Bunsen Committee and a Revision of the 'Conspiracy' of Sykes–Picot." *Middle Eastern Studies* 54 (4): 611–37.

Jones, Martin, and Gordon MacLeod. 2004. "Regional Spaces, Spaces of Regionalism: Territory, Insurgent Politics and the English Question." *Transactions of the Institute of British Geographers* 29:433–52.

Jones, Reece. 2012. *Border Walls: Security and the War on Terror in the United States, India, and Israel*. New York: Zed Books.

———. 2016. *Violent Borders*. New York: Verso.

———. 2021. *White Borders*. Boston: Beacon.

Kadercan, Burak. 2017. "Territorial Design and Grand Strategy in the Ottoman Empire." *Territory, Politics, Governance* 5 (2): 158–76.

Khalidi, Rashid. 1980. *British Policy toward Syria and Palestine 1906–1914: A Study of the Antecedents of the Hussein-McMahon Correspondence, the Sykes-Picot Agreement and the Balfour Declaration*. London: Ithaca.

———. 1997. *Palestinian Identity: The Construction of Modern National Consciousness*. New York: Columbia University Press.

———. 2004. *Resurrecting Empire*. Boston: Beacon.

Kieser, Hans-Lukas. 2019. *1914–1918 Online: International Encyclopedia of the First World War*. https://encyclopedia.1914-1918-online.net/home/.

Klauser, Francisco R. 2012. "Thinking through Territoriality: Introducing Claude Raffestin to Anglophone Sociospatial Theory." *Environment and Planning D: Society and Space* 30 (1): 106–20.

Kolers, Avery. 2009. *Land, Conflict, and Justice: A Political Theory of Territory*. New York: Cambridge University Press.

Krichker, Dina. 2021. "Making Sense of Borderscapes: Space, Imagination and Experience." *Geopolitics* 26:1224–42.

Kuch, Amelia. 2018. "Lessons from Tanzania's Historic Bid to Turn Refugees to Citizens." *Refugees Deeply*, February 22. https://deeply.thenewhumanitarian.org/refugees/community/2018/02/22/lessons-from-tanzanias-historic-bid-to-turn-refugees-to-citizens.

Labrador, Julián García, and José Ochoa. 2019. "Two Ontologies of Territory and a Legal Claim in the Ecuadorian Upper Amazon." *Journal of Political Ecology* 26:496–516.

Ledwith, Alison. 2014. *Za'atari: The Instant City*. Boston: Affordable Housing Institute.

Lefebvre, Henri. 1991. *The Production of Space*. Translated by Donald Nicholson-Smith. Malden, MA: Blackwell.

———. 2009. *State, Space, World: Selected Essays*. Translated by Gerald Moore, Neil Brenner, and Stuart Elden. Edited by Neil Brenner and Stuart Elden. Minneapolis: University of Minnesota Press.

Lenner, Katharina, and Lewis Turner. 2019. "Making Refugees Work? The Politics of Integrating Syrian Refugees into the Labor Market in Jordan." *Middle East Critique* 28 (1): 65–95.

Little, Adrian, and Nick Vaughan-Williams. 2017. "Stopping Boats, Saving Lives, Securing Subjects: Humanitarian Borders in Europe and Australia." *European Journal of International Relations* 23 (3): 533–56.

Lockman, Zachary. 2004. *Contending Visions of the Middle East*. Cambridge: Cambridge University Press.

Long, Katy. 2013. *The Point of No Return: Refugees, Rights, and Repatriation*. Oxford: Oxford University Press.

———. 2014. "Rethinking 'Durable' Solutions." In *Refugee and Forced Migration Studies*, edited by Elena Fiddian-Qasmiyeh, Gil Loescher, Katy Long, and Nando Sigona, 475–87. Oxford: Oxford University Press.

Lopez, Patricia J., Lisa Bhungalia, and Léonie S. Newhouse. 2015. "Geographies of Humanitarian Violence." *Environment and Planning A* 47:2232–39.

Magid, Aaron. 2017. "Amman's Refugee Waiting Game: The Time Bomb on Jordan's Border." *Foreign Affairs*, May 24, 2017.

Maier, Charles S. 2016. *Once within Borders: Territories of Power, Wealth, and Belonging since 1500*. Cambridge, MA: Harvard University Press.

Malkawi, Khetam. 2015. "Mafraq, Ramtha Population Doubled since Start of Syrian Crisis." *Jordan Times*, November 27, 2015.

Malkki, Liisa. 1992. "National Geographic: The Rooting of Peoples and the Territorialization of National Identity among Scholars and Refugees." *Cultural Anthropology* 7 (1): 24–44.

Mathema, Silva. 2018. "Refugees Thrive in America." Center for American Progress report, November 19, 2018. https://www.americanprogress.org/article/refugees-thrive-america/.

Marcu, Silvia. 2014. "Geography of Belonging: Nostalgic Attachment, Transnational Home and Global Mobility among Romanian Immigrants in Spain." *Journal of Cultural Geography* 31 (3): 326–45.

Marfleet, Philip. 2016. "Displacements of Memory." *Refuge* 32:7–17.

Marlowe, Jay. 2017. *Belonging and Transnational Refugee Settlement*. London: Routledge.

Marshood, Nabil. 2010. *Voices from the Camps: A People's History of Palestinian Refugees in Jordan, 2006*. Lanham, MD: University Press of America.

Martin, Diana. 2015. "From Spaces of Exception to 'Campscapes': Palestinian Refugee Camps and Informal Settlements in Beirut." *Political Geography* 44:9–18.

Martin, Diana, Claudio Minca, and Irit Katz. 2020. "Rethinking the Camp: On Spatial Technologies of Power and Resistance." *Progress in Human Geography* 44:743–68.

Mason, Victoria. 2007. "Ghurbah: Constructions and Negotiations of Home, Identity and Loyalty in the Palestinian Diaspora." In *Loyalties*, edited by Victoria Mason, 131–43. Salisbury, Australia: Griffin.

———. 2011. "The Im/mobilities of Iraqi Refugees in Jordan: Pan-Arabism, 'Hospitality' and the Figure of the 'Refugee.'" *Mobilities* 6 (3): 353–73.

Massad, Joseph. 2001. *Colonial Effects: The Making of National Identity in Jordan*. New York: Columbia University Press.

Massey, Doreen. 2005. *For Space*. London: Sage.

McCarthy, Niall. 2017. "Lebanon Still Has Hosts the Most Refugees per Capita by Far." *Forbes*, April 3, 2017. https://www.forbes.com/sites/niallmccarthy/2017/04/03/lebanon-still-has-hosts-the-most-refugees-per-capita-by-far-infographic/.

McConnachie, Kirsten. 2016. "Camps of Containment: A Genealogy of the Refugee Camp." *Humanity* 7 (3): 397–412.

Milton-Edwards, Beverley, and Peter Hinchcliffe. 2009. *Jordan: A Hashemite Legacy*. 2nd ed. New York: Routledge.

Ministry of Planning and International Cooperation of the Hashemite Kingdom of Jordan. 2017. *Jordan Response Plan for the Syria Crisis 2018–2020, Executive Summary*. Amman: Ministry of Planning and International Cooperation.

Modeliranje, Dasimetrocno, Mileva Samardzic-Petrovic, Nikola Krunic, and Milan Kilibarda. 2013. "Dasymetric Modelling of Population Dynamics in Urban Areas." *Geodetski Vestnik* 57:777–92.

Mohanty, Chandra Talpade. 2003. "Feminist Encounters: Locating the Politics of Experience." In *Feminist Theory Reader*, edited by Carole McCann and Seung-Kyung Kim, 460–71. New York: Routledge.

Molnar, Petra. 2017. "Discretion to Deport: Intersections between Health and Detention of Syrian Refugees in Jordan." *Refuge* 33 (2): 18–31.

Monsutti, Alessandro. 2010. "The Transnational Turn in Migration Studies and the Afghan Social Networks." In *Dispossession and Displacement: Forced Migration in the Middle East and North Africa*, edited by Dawn Chatty and Bill Finlayson, 45–67. Oxford: Oxford University Press.

Mountz, Alison. 2009. "Migration." In *Key Concepts in Political Geography*, edited by Carolyn Gallaher, Carl Dahlman, Mary Gilmartin, and Alison Mountz. Los Angeles: Sage.

Murphy, Alexander. 2013. "Territory's Continuing Allure." *Annals of the Association of American Geographers* 103 (5): 1212–26.

Nassar, Jessy, and Nora Stel. 2019. "Lebanon's Response to the Syrian Refugee Crisis—Institutional Ambiguity as a Governance Strategy." *Political Geography* 70:44–54.

Natali, Denise. 2010. *The Kurdish Quasi-State: Development and Dependency in Post-Gulf War Iraq*. Syracuse, NY: Syracuse University Press.

Naylor, Lindsay, Michelle Daigle, Sofia Zaragocin, Margaret Marietta Ramírez, and Mary Gilmartin. 2018. "Interventions: Bringing the Decolonial to Political Geography." *Political Geography* 66:199–209.

Newman, David. 2006. "The Lines That Continue to Separate Us: Borders in Our 'Borderless' World." *Progress in Human Geography* 30 (2): 143–61.

Oesch, Lucas. 2016. "The Refugee Camp as a Space of Multiple Ambiguities and Subjectivities." *Political Geography* 60:110–20.

Ong, Aihwa. 1999. *Flexible Citizenship: The Cultural Logics of Transnationality*. Durham, NC. Duke University Press.

Osiander, Andreas. 2001. "Sovereignty, International Relations, and the Westphalian Myth." *International Organization* 55 (2): 251–87.

Painter, Joe. 2008. "Cartographic Anxiety and the Search for Regionality." *Environment and Planning A* 40:342–61.

———. 2010. "Rethinking Territory." *Antipode* 42 (5): 1090–118.

Pappe, Ilan. 1994. "Jordan between Hashemite and Palestinian Identity." In *Jordan in the Middle East: The Making of a Pivotal State 1948–1988*, edited by Joseph Vevo and Ilan Pappe. Portland, OR: Frank Cass.

Peteet, Julie. 1995. "Transforming Trust: Dispossession and Empowerment among Palestinian Refugees." In *Mistrusting Refugees*, edited by E. Valentine Daniel and John Chr. Knudsen, 168–86. Berkeley: University of California Press.

Peters, Kimberly, Philip Steinberg, and Elain Stratford. 2018. "Introduction." In *Terra beyond Territory*, edited by Kimberly Peters, Philip Stienberg, and Elain Stratford, 1–13. Lanham, MD: Rowman & Littlefield.

Philipp, Thomas. 2004. "Bilad al-Sham in the Modern Period: Integration into the Ottoman Empire and New Relations with Europe." *Arabica* 51:401–18.

Pitcher, Donald. 1968, 1972. *An Historical Geography of the Ottoman Empire from Earliest Times to the End of the 16th Century*. Leiden, Netherlands: Brill.

Porath, Y. 1984. "Abdallah's Greater Syria Programme." *Middle Eastern Studies* 20 (2): 172–89.

Provence, Michael 2017. *The Last Ottoman Generation and the Making of the Modern Middle East*. Cambridge: Cambridge University Press.

Quiquivix, Linda. 2014. "Art of War, Art of Resistance: Palestinian Counter-cartography on Google Earth." *Annals of the Association of American Geographers* 104 (3): 444–59.

Radcliffe, Sarah. 2017. "Decolonising Geographical Knowledges." *Transactions of the British Institute of Geographers* 42 (3): 329–33.

Raffestin, Claude. 1984. "Territoriality: A Reflection of the Discrepancies between the Organization of Space and Individual Liberty." *International Political Science Review* 5 (2): 139–46.

———. 2007. "Could Foucault Have Revolutionized Geography?" In *Space, Knowledge, and Power: Foucault and Geography*, edited by Jeremy W. Crampton and Stuart Elden, 129–37. Burlington, VT: Ashgate.

———. 2012. "Space, Territory, and Territoriality." Translated by S. A. Butler. *Environment and Planning D: Society and Space* 30 (1): 121–41.

Rajaram, Prem Kumar. 2002. "Humanitarianism and Representations of the Refugee." *Journal of Refugee Studies* 15 (3): 247–64.

Ramadan, Adam. 2012. "Spatialising the Refugee Camp." *Transactions of the Institute of British Geographers* 38:65–77.

Ramadan, Adam, and Sara Fregonese. 2017. "Hybrid Sovereignty and the State of Exception in the Palestinian Refugee Camps in Lebanon." *Annals of the Association of American Geographers* 107 (4): 949–63.

Ramaswamy, Sumathi. 2010. *The Goddess and the Nation: Mapping Mother India*. Durham, NC: Duke University Press.

Ratzel, Fredrich. 1896. "The Territorial Growth of States." *Scottish Geographical Magazine* 12 (7): 351–61.

Robertson, Sean, Simon Okpakok, and Gita Ljubicic. Forthcoming. "Territorializing Piquhiit in Uqsuqtuuq (Gjoa Haven, Nunavut, Canada): Negotiating Homeland through an Inuit Normative System." *Territory, Politics, Governance*. Published ahead of print, Nov. 5, 2020. https://doi.org/10.1080/21622671.2020.1837664.

Robson, Laura. 2017. "Refugees and the Case for International Authority in the Middle East: The League of Nations and the United Nations Relief and Works Agency for Palestinian Refugees in the Near East Compared." *International Journal of Middle East Studies* 49:625–44.

Rutinwa, Bonaventure. 2002. "Prima Facie Status and Refugee Protection." UNHCR Evaluation and Policy Analysis Unit, working paper no. 69. https://www.unhcr.org/en-us/research/working/3db9636c4/prima-facie-status-refugee-protection-bonaventure-rutinwa.html.

Ryan, Curtis. 2010. "'We Are All Jordan' . . . but Who Is We?" *Middle East Research and Information Project*. https://merip.org/2010/07/we-are-all-jordan-but-who-is-we/.

———. 2011. "Identity Politics, Reform, and Protest in Jordan." *Studies in Ethnicity and Nationalism* 11 (3): 564–78.

Sack, Robert. 1986. *Human Territoriality*. New York: Cambridge University Press.

Sadek, George. 2013. *Legal Status of Refugees: Egypt, Jordan, Lebanon, and Iraq*. Washington, DC: Law Library of Congress, Global Legal Research Center.

Said, Edward. 1978. *Orientalism*. New York: Vintage Books.

———. 1992. *The Question of Palestine*. New York: Vintage Books.

———. 1993. *Culture and Imperialism*. New York: Vintage Books.

Saliba, Issam. 2016. "Jordan." In *Refugee Law and Policy in Selected Countries*, 213–15. Washington, DC: Library of Congress. https://irp.fas.org/congress/2016_rpt/lloc-refugee.pdf.

Sanyal, Romola. 2014. "Refugees: The Work of Exile: Protracted Refugee Situations and the New Palestinian Normal." In *The Post-Conflict Environment: Investigation and Critique*, edited by Daniel Bertrand Monk and Jacob Mundy, 135–57. Ann Arbor, MI: University of Michigan Press.

———. 2017. "A No-Camp Policy: Interrogating Informal Settlements in Lebanon." *Geoforum* 84:117–25.

———. 2018. "Managing through Ad Hoc Measures: Syrian Refugees and the Politics of Waiting in Lebanon." *Political Geography* 66:67–75.

Sassen, Saskia. 2008. *Territory, Authority, Rights: From Medieval to Global Assemblages*. Princeton, NJ: Princeton University Press.

Schofield, Richard. 2018. "International Boundaries and Borderlands in the Middle East: Balancing Context, Exceptionalism and Representation." *Geopolitics* 23 (3): 608–31.

Schwedler, Jillian. 2022. *Protesting Jordan: Geographies of Power and Dissent*. Stanford, CA: Stanford University Press.

Shami, Seteney. 1996. "Transnationalism and Refugee Studies: Rethinking Forced Migration and Identity in the Middle East." *Journal of Refugee Studies* 9 (1): 3–26.

Shiblak, Abbas. 1996. "Residency Status and Civil Rights of Palestinian Refugees in Arab Countries." *Journal of Palestine Studies* 25 (3): 36–45.

Simpson, Charles. 2018. "Competing Security and Humanitarian Imperatives in the Berm." *Forced Migration Review* 57:15–18.

Singh, Ashika L. 2020. "Arendt in the Refugee Camp: The Political Agency of World-Building." *Political Geography* 77:102149. https://doi.org/10.1016/j.polgeo.2020.102149.

Sluglett, Peter. 2010. "Introduction," in *Syria and Bilad al-Sham under Ottoman Rule*. Leiden, Netherlands: Brill.

Smith, Neil. 2003. *American Empire: Roosevelt's Geographer and the Prelude to Globalization*. Berkeley: University of California Press.

Smith, Sara. 2013. "'In the Past We Ate from the Same Plate': Memory and the Border in Leh, Ladakh." *Political Geography* 35:47–59.

Soja, Edward. 1996. *Third Space*. Malden, MA: Blackwell.

Sparke, Matthew. 1998. "A Map That Roared and an Original Atlas: Canada, Cartography, and the Narration of Nation." *Annals of the Association of American Geography* 88:463–95.

Spivak, Gayatri Chakravorty. 1988. "Can the Subaltern Speak?" In *Marxism and the Interpretation of Culture*, edited by C. Nelson and L. Grossberg. Urbana: University of Illinois Press.

Squire, Vicki. 2020. *Europe's Migration Crisis: Border Deaths and Human Dignity*. Cambridge: Cambridge University Press.

Steinberg, Philip. 2009. "Sovereignty, Territory, and the Mapping of Mobility: A View from the Outside." *Annals of the Association of American Geographers* 99 (3): 467–95.

Stevens, Dallal. 2013. "Legal Status, Labelling, and Protection: the Case of Iraqi 'Refugees' in Jordan." *International Journal of Refugee Law* 25 (1): 1–38.

Su, Alice. 2013. "How Do You Rank Refugees?" *The Atlantic*, November 22, 2013.

Susser, Asher. 1994. "Jordan, the PLO and the Palestine Question." In *Jordan and the Middle East: The Making of a Pivotal State*, edited by Joseph Nevo and Ilan Pappe, 211–28. Portland, OR: Frank Cass.

Sylvester, Christine. 2012. "War Experiences/War Practices/War Theory." *Millennium: Journal of International Studies* 40 (3): 483–503.

Syrian Needs Analysis Project. 2013. "Legal Status of Individuals Fleeing Syria." https://reliefweb.int/report/syrian-arab-republic/legal-status-individuals-fleeing-syria-syria-needs-analysis-project-june.

Takkenberg, Lex. 2010. "UNRWA and the Palestinian Refugees after Sixty Years: Some Reflections" *Refugee Survey Quarterly* 28 (2/3): 253–59.

Tal, Lawrence. 1993. "Is Jordan Doomed?" *Foreign Affairs* 72 (5): 45–58.

Tanenbaum, Jan Karl. 1978. “France and the Arab Middle East, 1914–1920.” *Transactions of the American Philosophical Society* 68 (7): 1–50.

Tiltnes, Åge A., and Huafeng Zhang. 2013. “Progress, Challenges, Diversity: Insights into the Socio-economic Conditions of Palestinian Refugees in Jordan.” Fafo. https://www.unrwa.org/sites/default/files/insights_into_the_socio-economic_conditions_of_palestinian_refugees_in_jordan.pdf.

Tobin, Sarah A., and Madeline Otis Campbell. 2016. “NGO Governance and Syrian Refugee ‘Subjects’ in Jordan.” *Middle East Report* 278. https://merip.org/2016/04/ngo-governance-and-syrian-refugee-subjects-in-jordan/.

Tolia-Kelly, Divya. 2004. “Locating Processes of Identification: Studying the Precipitates of Re-memory through Artefacts in the British Asian Home.” *Transactions of the Institute of British Geographers* 29: 14–29.

United Nations (UN). 1950. “Convention Relating to the Status of Refugees.” https://www.unhcr.org/en-us/5d9ed32b4.

United Nations High Commissioner for Refugees (UNHCR). 2013a. “UNHCR Policy on Alternatives to Camps.” https://www.unhcr.org/en-us/protection/statelessness/5422b8f09/unhcr-policy-alternatives-camps.html.

———. March 2013b. “Universal Periodic Review: Jordan.” https://www.refworld.org/country,,UNHCR,,JOR,,513d90172,0.html.

———. 2015. “Global Trends: Forced Displacement in 2014.” https://www.unhcr.org/en-us/statistics/country/556725e69/unhcr-global-trends-2014.html.

———. 2017. “Global Trends: Forced Displacement in 2017.” https://www.unhcr.org/globaltrends2017/.

———. 2018a. “UNHCR Fact Sheet: Jordan.” https://reporting.unhcr.org/sites/default/files/UNHCR%20Jordan%20Fact%20Sheet%20-%20June%202018.pdf.

———. 2018b. “UNHCR Service Guide.” https://www.unhcr.org/jo/wp-content/uploads/sites/60/2018/08/WEB-FINAL_Service-Guide-August2018_ENG-HighRes.pdf.

———. 2018c. “Za'atari Refugee Camp: Factsheet, February 2018.” https://reliefweb.int/report/jordan/zaatari-refugee-camp-factsheet-february-2018.

———. 2019. “UNHCR Fact Sheet: Jordan.” https://data.unhcr.org/en/documents/details/69826.

———. January 2021. “Jordan: Zaatari Refugee Camp.” https://data.unhcr.org/en/documents/details/85014.

———. 2022. “Situation Syria Regional Refugee Response.” https://data.unhcr.org/en/situations/syria/location/36.

United Nations Relief and Works Agency for Palestinians in the Near East (UNRWA). 2022. “Palestine Refugees.” https://www.unrwa.org/palestine-refugees.

Usher, Mark. 2020. “Territory Incognita.” *Progress in Human Geography* 44 (6): 1019–46.

Van Hear, Nicholas 2003. “From Durable Solutions to Transnational Relations: Home and Exile among Refugee Diasporas.” UNHCR Evaluation and Policy Analysis Unit working paper no. 83. https://www.unhcr.org/research/working/3e71f8984/durable-solutions-transnational-relations-home-exile-among-refugee-diasporas.html.

Wastl-Walter, Doris, and Lynn Staeheli. 2004. “Territory, Territoriality, and Boundaries.” In *Mapping Women, Mapping Politics*, edited by Lynn Staeheli, Eleonore Kofman, and Linda Peake, 141–51. New York: Routledge.

Watkins, Josh. 2020. “Irregular Migration, Borders, and the Moral Geographies of Migration Management.” *Environment and Planning C: Politics and Space* 38 (6): 1108–27.

Wolford, Wendy. 2004. "This Land Is Ours Now: Spatial Imaginaries and the Struggle for Land in Brazil." *Annals of the Association of American Geographers* 94 (2): 409–24.

Woroniecka-Krzyzanowska, Dorota. 2017. "The Right to the Camp: Spatial Politics of Protracted Encampment in the West Bank." *Political Geography* 61: 60–69.

Wright, Sarah. 2015. "More-Than-Human, Emergent Belongings: A Weak Theory Approach." *Progress in Human Geography* 34 (4): 391–411.

Yassin-Kassab, Robin, and Leila Al-Shami. 2016. *Burning Country: Syrians in Revolution and War*. London: Pluto.

Yuval-Davis, Nira. 2006. "Belonging and the Politics of Belonging." *Patterns of Prejudice* 40 (3): 197–214.

Zaman, Tahir. 2016. "A Right to Neighbourhood: Rethinking Islamic Narratives and Practices of Hospitality in a Sedentarist World." In *The Refugee Crisis and Religion: Secularism, Security and Hospitality in Question*, edited by L. Mavelli and E. Wilson. Lanham, MD: Rowman & Littlefield.

Zaragocin, Sofia, and Martina Angela Caretta. 2020. "Cuerpo-Territorio: A Decolonial Feminist Geographical Method for the Study of Embodiment." *Annals of the American Association of Geographers* 111 (5): 1503–18.

Index

Page numbers followed by "f" refer to figures.

Printed and bound by CPI Group (UK) Ltd, Croydon, CR0 4YY

27/01/2025

14633043-0002